Life with Larry Thompson

Life with Larry Thompson

Columns which appeared in the Miami Herald,

Selected by

Larry's "Good Wife"

Penny Thompson

WAKE - BROOK HOUSE

 For information address Wake-Brook House, Book Publishers, 960 NW 53rd Street, Fort Lauderdale, Florida, 33309.

International Standard Book Number:

O-87482-047-2

Printed in the United States of America.

First Edition.

Acknowledgment is made to the Miami Herald for permission to reprint the material used in this book, which appeared in the columns of that newspaper over the past two decades.

Contents

Foreword

By Carl Thompson

Rarely does one find two people that are both successful personalities as well as dedicated parents.

Larry and Penny Thompson were such people.

I know, I was their son.

While tenaciously cornering a suspicious subject for a story as a prize-winning Miami Herald reporter, my father also would be painstakingly mulling in his head ways to love us children and his "good wife."

I remember him often coming home, bubbling with excitement as he proudly presented my mother with an oversized dress or fancy jewelry, which she never desired, but none the less made him happy.

The hands that madly pounded out a story for deadline at the newspaper would also gently caress our foreheads as we lay in bed with a fever.

The role was changed as Dad became more and more ill with emphysema over the years. His long and frequent stays in the hospital became more serious.

We children now would hold his hand and tell him all about our day, while praying silently he would live.

Although in severe pain, he would always manage a weak smile and his eyes would convey his most important message – that he loved us and was proud of his children.

The doctors said this did him more good than all the medicine they gave him, for he would "melt like butter" when his wife or children "warmed up" to him, as he once secretly wrote on a piece of paper.

He loved life and fought hard to keep it. He enjoyed laughing at his blunders and sharing it with his readers. He cherished his family, bringing them to church on Sunday and watching his children grow.

He was a rare man, and in my eyes a great one.

Our mother surely was a godsend for his life.

During his last years of partial blindness, she patiently and gently led him by the arm through the streets of New York City and the countries of Western Europe while on vacation.

She was his nurse at home, carefully preparing a multitude of pills and medicines for him to take before writing his column each day.

She was his notebook – remembering things he had forgotten or couldn't pronounce.

She was his sweetheart, she was his good wife.

As we grew older, we learned how Mother had given up a glamorous career as a woman flier and aviation promoter to become the constant companion of Dad and we three children.

During the 40's she spearheaded the drive to name the Amelia Earhart Field at Opa-Locka in honor of the famous aviatrix. She also was busy organizing international air shows for women and publishing her own aviation newspaper which was distributed throughout the Southeastern United States.

While in the Civil Air Patrol, Mother flew over the Gulf of Mexico, searching for German submarines during World War II.

The Annual Twin's Easter Parade through downtown Miami was started some 21 years ago by Mother. It was the result of her desire to give her children a special kind of birthday party – with a parade.

After Father's death in 1973, Mother's dream was to publish this book of his best and most popular columns. For more than a year she labored through some 27 years of his writings and carefully selected the columns which appear in this book.

Mother didn't live to see her dream come true. She died September 22, 1975 of acute leukemia, before the book was printed.

The world is a more empty place without Larry and Penny Thompson – but the memories of our happy days with them will live on.

Classified Section
OH, BOY! THINK O' THE SWELL READIN' I BEEN MISSIN'!
LATELY I HAD TO PRETEND TO READ OFFICE BULLETINS AND OFTEN SMILED WHEN SMILES WEREN'T IN ORDER.
DIET
WONDER IF THEY HAVE REDUCIN' LENSES?

Florida by Moonglow

FLORIDA BY MOONGLOW

I fell in love with South Florida years ago when I stepped off the train from New York, carrying my top coat and a suit case full of heavy clothes and even galoshes.

I stepped off the train into the most ramshackle, ugliest depot in the world, but once I got out of that dump heap, I stepped into a world of sunshine and beauty, even though it was December.

There were flowers blooming and the grass was green and the air was fresh and mixed with salt and the scent of tropical growth, and people were in shirt sleeves, and the snows and slush of Gotham were far behind.

I fell in love with South Florida then, and I haven't fallen out yet. But, just as you get used to your parents or your children or your wife or the other precious things of life, you sometimes take a thing or a place or a person for granted.

Every now and then I am jolted into an appreciation of why I am enamoured of this part of the earth.

The other night we had a full moon. I didn't know this when I left the house to walk to the corner drug store, but as soon as I stepped off the front stoop I was flooded by the moon's bright light.

Nowhere in the world can the moon be so big and so close to the earth as here. I haven't been everywhere, but I don't see how that moon could be any bigger and get any closer without having a collision.

It is huge and it is bright, and it is beautiful and mellow. They could turn off the street lights when it is shining at full face, but of course the best way to appreciate it would not be in the city but walking along a beach or in a flower garden or a grove.

But even in a city, it's a tremendous sight, one that makes you stop and look to the heavens in appreciation and to look around you and see familiar surroundings in proper perspective.

One of our neighbors has flowers planted between the sidewalk and the curb. They are white flowers and in the strong moonlight, they looked as if they had been painted with luminous silver paint. In South Florida, flowers bloom all night, and in the full moon the colors of all except the deepest hues can be seen as clearly as by daylight, but with a different, a more exotic appearance.

The next morning, I noticed that the bougainvillea at the side of our house was busting out all over with brilliant purple blossoms, and on the way to town I saw the Golden Shower trees in the center of Brickell Ave. were covered with magnificent yellow flowers.

As I said, I've been in love with South Florida for a long time. Every now and then when I take a close look at her, I know why.

Calypso Growing Old

One of my most pleasant memories is of a visit to Nassau years ago with my wife, then a blushing bride.

One night I hired a sailboat for a pittance and for another pittance I hired a native boy who was playing a guitar and singing calypso on the wharf.

The moon, as I remember it, was the most romantic in its history and it painted phosphorescent pictures as our sailboat man maneuvered us through the harbor and into adjacent waters. He and our music man joined in singing calypso, and people from expensive yachts and big liners looked out in envy as our small vessel with its white sails skipped over the moonlit waves leaving a wake of haunting melody.

My wife, who was too young then to know better, thought I was wonderful. I was old enough to know better, but I shared her opinion, since the cruise had turned out so well.

As a result, my wife and I became avid calypso fans. Maybe we really were thinking of that sailboat ride, but the music to us seemed plaintive and romantic, even if the words were robust or comical.

We bought many calypso records and played them often, and then we made a pair of babies (to borrow a calypso term) bought a TV set, and sort of forgot the whole thing.

Then there were three babies and they discovered Captain Kangaroo, Quick Draw McGraw, Lassie plus many less worthwhile attractions on TV, and we never did get around to playing our calypso records any more.

The years went by and recently our TV quit making pictures.

To entertain our children, who think it is a rule that they must fight one another if there isn't something on TV, we dug out some discs (you know, like the disc jockeys ride) and told them they could use the record player.

One of the first records they played was a calypso, and they instantly fell in love with it.

My wife and I were delighted, not only at hearing the old songs again, but to find that our children shared our affection for such music.

Happily we listened as a singer said, "Brown-skin girl go home and mind baby," as he brayed like a donkey in the "Jackass Serenade," and informed his date that "My name is Morgan, but It Ain't J.P.," or as he implored Liza to come back and "wipe the tear from me eye."

But by the 10th time around, we began to notice that the volume on the record player was very high. I imagine our neighbors already had noticed this.

My wife broke first. She looked as if she felt guilty for admitting it, but after the 15th time she commented she was beginning to get tired of that record. She tried to convince the children that Bing Crosby's "White Christmas" would be a nice record for a change, or to at least try Judy Garland's "Wizard of Oz" recording.

All we got was more about every time the singer was away from Liza "water come to me eye, so come back Liza, come back girl, wipe the tear from me eye."

I began to wish that guy would buy a handkerchief and quit crying for Liza.

He was braying like a donkey and crying like a baby at all hours, especially in the early morning when I was trying to get another half hour sleep.

Not only did I get so that I hated Liza and her boy friend, but I couldn't stand Morgan, J.P. or otherwise, and I was fed up with brown-skinned girls who ought to be home minding babies.

The worst of it was that I heard these songs so much I began to sing them.

I have decided that record must go, even if I must shatter it to pieces in the dead of night. Just today the editor stopped to ask me a question and, the calypso on my mind, I brayed the chorus from the "Jackass Serenade."

I needed somebody to wipe the tear from me eye when he walked away without waiting for an explanation.

Memories? They're Made of Bikes and Kisses and Sleeping Children

They say that when you are drowning your life flashes through your mind. In that brief time, even at a fast pace, you wouldn't get a complete show, from birth to demise. The mental performance would have to consist of the highlights of your mortal existence.

And just what are the highlights? What are the things that have impressed you the most during your tenure on this earth?

These wouldn't necessarily be the most important or significant events. For instance, your birth is the most important part of your life, but you don't remember anything about it. Even when drowning, I doubt if you'd have a scene devoted to your debut.

The first day in school was an event of great importance and one that should be engraved in your memory. Maybe I'm weak minded, but I must admit I can't remember that big day, except what my mother has told me about it.

In fact, I think we imagine most of our memories of childhood. We have heard our parents and relatives tell stories so often that we think we remember them, but we're really remembering their stories.

One of the first clear, independent recollections I have of a Real Important event is the day I got my first bicycle. It was second-hand and it was much too big for me, but I had been asking for a bike for so long, and riding my friends' bikes for so long when I could borrow or sneak off with them, that this was a memorable occasion.

Another highlight of about the same time of life – it may have been a few years before or after I got the bike – was sad. That was the day I came home from school and Trixie, my golden and white collie, didn't meet me at the corner. I found her in the back yard, dead. She had been poisoned.

Even though I thought it was unmanly to cry, I cried uncontrollably. She had been my best friend.

We buried her between the garage and the rose arbor, and my father helped me make a headstone from a big sandstone.

Another highlight was the first time I fell in love. I no longer remember the girl's name, but I can see her in my mind's eye, and she was the most wonderful girl who ever lived. I walked home with her once – probably from Epworth League – and we talked for a long time on her front porch. When I left I leaned over and lightly kissed her on a cheek.

I ran home, leaping to grab leaves off low-hanging boughs, shouting and singing, and deliriously happy with life and young love. The next day I was embarrassed to look at my love, for I was overcome with my brashness of the night before, and I don't even know if I ever had a real date with her after, but for a time I savored the supreme bliss of romance.

Winning the Payne County championship wrestling title was another highlight of my adolescent days, as well as the peak of my athletic career. I doubt if I have ever been prouder than the day the letter sweaters arrived and I walked the streets in mine, pretending it was nothing, but strutting like a peacock.

Two main highlights concern the newspaper business. One was seeing my first story in print. I was in the ninth grade and got a non-paying job as reporter on the college paper that summer. That first story was only two paragraphs long, but I still have it in a scrap book. The other journalistic highlight was seeing my first by-line. That was the following summer, when I worked for nothing on the home town paper. I still like to see my by-line, but none has ever given me such a thrill.

Seeing the first edition of the college newspaper roll off the presses when I was editor was another highlight that belongs in the professional category.

Among the most indelible highlights were the first views of my new-born children, and the face of my wife when she saw them. There is no radiance like that of a mother with her baby.

Any many highlights blend into that one – that of tiptoeing into a bedroom to see the sleeping children. These are rare moments, for it is only then that mortal man can see the face of angels.

Oh, there are many more highlights that I would remember. That would have to be a slow drowning to get in the entire show.

Only His Eyes Tell . . .

Yesterday was Father's Day.

In Stillwater, Okla., the doorbell rang and a young man handed my mother a telegram.

"It's for you, Carl," she said and handed it to the gaunt ghost of a man seated in the big chair watching television.

A stranger might have thought that he showed but little response. He did not reach for the telegram. He turned his eyes only slightly at the message.

But my mother saw his eyes light up with that old intensity and she knew he had heard. She knew, too, that he was too crippled to make any more demonstration of the feelings that were surging inside his once-proud body.

She opened the telegram and read it to him. As they had both known it would be, it was a Father's Day greeting from their son in Florida, the one of their four children who always forgot to buy a card on time to mail it on any special occasion. But it was from their only begotten son and the old man smiled a crooked smile of pride and happiness with the half of his lips that are not paralyzed, for even in the days and years of his helpless agony he is proud of his boy and happy at any word from him.

He is proud of his three daughters, too, and of his grandchildren and his great-grandchild, and of his wife who now is his nurse as well as his wife. He is a family man, a man of deep love and loyalty for his own, a man who was stern and gentle, a man who set a perfect example for his children to follow, yet was understanding when they faltered along the way.

He was beset by many ailments during his life, but always he was a strong man. We never could think of him as otherwise, for always he was full of courage and plans for our future. He believed in God and it never occurred to him that God would not care for him and his loved ones. He made God's task easy, for he cared for himself and his own so well that there was little for the Lord to do but nod approval.

I do not mean that he was rich in material things, for he made a professor's pay, which was not sufficient for lavish material things for a large family. But he was rich in every important phase of living, and he passed on those riches to his children, for he always was a man who shared.

Of course, he did not confine his influence to his family.

As a Sunday School teacher and superintendent, as a Boy Scout leader, as a Rotary Club official, as a teacher and as a public speaker his influence was far flung. No student attended one of his classes without profiting from it, regardless of how much he learned from the book. His students became his admirers and wherever he went he left more friends behind him, for he was a rare man without malice or cruelty or pettiness, a good man who was not sanctimonious, a gentle man who was not prissy, a strong man who was not arrogant.

He was a man who liked to laugh and he liked to talk. There probably wasn't a high school in the state of Oklahoma where he hadn't addressed the graduating class, not a 4-H or FFA club he hadn't addressed, hardly a state convention at which he hadn't spoken. When he retired from college, for a time he traveled about the country, from San Francisco to Miami, addressing national conventions and organizations.

He liked people and he liked to talk to people. He could never get away from church or a meeting for at least half an hour after adjournment, because he had to stop and talk to so many folks.

Then, several years ago, he had a heart attack and another, and he fought them off and was well, only to be crippled by the most severe stroke of all. And now, for several years, he can barely walk, though he forces himself to get about the house unaided, he cannot write a note, and he cannot speak a word.

People have not forgotten him. Friends still visit him, even after such a prolonged confinement, but they must talk to him, and he can only listen.

Only his eyes could show his reaction when a telegram was read to him yesterday. And he was grateful for such a small token from a boy who owes him so much.

God bless you, Papa, not only on Father's Day, but every day, as a man who is both humble and great, and is dearly beloved by many.

Two Very Proud Dads

My three sisters and I went from our homes in Oklahoma, Colorado, New Jersey and Florida to our hometown to observe the golden wedding anniversary of our parents. It had been many years since we had all been together. The anniversary was on a Saturday and it was assumed by all of us, without discussion, that we would go to church the next morning as a family group, just as we had each Sunday when we were children.

We walked down the aisle together in single file and we took up most of one pew near the front and when the services were over, it seemed that everyone in the congregation came over to speak to the Carl Thompsons and their kids.

The thing that most of these people mentioned was how proud my father looked, and I had noticed it myself for his pride was almost shining from him, it was so intense.

He never saw us all together again.

Last Sunday my own family went to church. True, this wasn't a reunion, for we all live in the same home and see one another every day. But since my children became young people we almost never go anyplace as a group. From the last year of junior high school on, their interests have extended beyond the family circle and each one has more and more activities away from home with his or her own friends.

My good wife and I are not too happy about this, but it is a fact of life that as children become mature, their social and extra-curricular activities no longer are arranged at home or include the family. This is as it must be, and should be.

It has been a long time since we could suddenly announce that the family was going on a picnic or to the beach or to visit a friend. If such an announcement were made now, the boy would have football practice or have to bag groceries, the oldest girl would have a meeting or date, the youngest girl would have a baby-sitting job or a school party.

I have wandered. I had intended only to tell about last Sunday morning.

One of the family traditions I was determined to preserve was our group attendance at Easter church services. I was in the hospital this Easter and although the rest of the family attended, I felt cheated. And so I announced that we would all go together on Mother's Day.

You could tell it was Mother's Day by listening.

"Mommy, tell Carl he has to let me in the bathroom! He's been in there half an hour!"

"Mommy! Tell Evellyn she can't wear my stockings!"

"Mommy! Where's the brush?" "Mommy, where are my best shoes?" "Penny, what happened to my cufflinks?"

And poor mother ran from hither to thither and found everybody's everythings and told people to eat and to hurry and then finally she had everyone dressed, fed and shining and quickly turned herself into a cool and calm looking well-dressed lady and off we went, late as usual.

And we marched down the aisle, regrettably tardy but conspicuous.

We almost filled a pew and I thought to myself as I saw my overworked but attractive wife, my two blonde and beautiful daughters, my towering son, uncomfortable but handsome in a coat, that these were mine, my own, this was my family, and I hoped that everybody got a good look at us, all of us together.

That is when my mind skipped from that church to another church far away and years before.

At last, I fully understood my father's pride, for now it was my own.

WHAT IS IT? A BOY OR A GIRL?
WAITING ROOM
BOTH-
ONE OF EACH, YOU LUCKY BOY!
DOUBLE TROUBLE
QUIN HALL

DIAPER SERVICE
BABY CARE

Joys of a New Mother

A letter to my daughter who has just become a mother for the first time.

You have entered the wonderful and enchanted world of motherhood, the ultimate goal of all women.

Regardless of station or riches, no woman's life is complete unless she has been a mother and no satisfaction is as deep as that which you enjoyed when you painfully gave birth to a new life.

You told me, in a weak and happy voice, that your daughter was the most beautiful baby you had ever seen.

Naturally you would think so. Every mother thinks that of her children, especially the first born.

To you, the red and wrinkled face, the wizened and helpless body of flesh and soft bone, were beautiful and you were thankful that you were permitted to give life to this creation.

And of course you should be thankful, for you not only have performed the primary function of a woman in reproducing your race, but you have brought into the world a human being who will be a source of great delight, as well as great worry; a human being you can help mold, just as you helped create her; a girl who will grow up and take her place in the society of humanity; a woman who will mature and pass on to future generations the character which you helped place in her bloodstream before her birth and will help instill in her brain in her maturing years.

In her, you have your memorial, your insurance of immortality, your link to posterity, your promise of eternity.

In most of the first year, the baby will remain helpless, and you will enjoy one of the greatest thrills of motherhood. Your child can eat only from you or because of you.

Your child is completely dependent on you for its existence, just as she was during the months she was awaiting birth.

Despite the disagreeable parts of the duty, you will thrill to the knowledge that this tiny mortal depends entirely on you for all her wants, and you will know deep happiness as you lull your child to sleep, feed your child when there is hunger, and treat your child when there is sickness.

Once your child learns to talk and walk, she is becoming an individual and you will realize that she is growing away from you.

But those are delightful years, too, even when your child climbs on the chairs and knocks down the breakable things from the shelves and when you despair of ever having the house clean again or having a moment you can call your own.

The years of 2 and 3 are among the most precious of the years of a child, although all of them you will treasure in your memory.

But as the child struggles to think and to express thoughts, you are seeing a miracle happening and you cannot help but glory in the fact that this is your baby who is learning how not to be a baby.

You will cry many times at the minor transgressions of your growing child and you will wonder at times if your worries will ever end, but there will be many moments when you are so well repaid for your labors and your worries that you will rejoice at your good fortune.

These will be when you hear your child saying prayers at night; or when you cure a hurt finger with a kiss; or when a child crawls into your lap and becomes a baby again; when your child runs to meet you or puts her arms out to greet you, or plants a wet kiss on your cheek.

And there will be such a time when you tiptoe into the child's room at the end of a day when she has been in nothing but trouble, a day of noise, dirt, and hurt, and you see her sleeping with the beauty and innocence of an angel – or a baby.

The child will leave you some day, as she should, and you may feel neglected as the years accumulate on your shoulders.

But the time will come when your child will have her own child, as you, my daughter, have done, and then, in that distant year, you will realize why you now are so blessed that you have become a mother.

You will know, then at last, that when an infant is placed in the arms of a mother, she is holding the destiny of the human race in her Arms.

Letter to a Son

Dear Son:

You are now a man, and that means that your father is supposed to have a man-to-man talk with you. Your character probably is too firmly moulded now to be affected by advice, but it makes the father feel righteous to offer it.

Naturally I hope you are a success, even if you aren't president.

I want you to be a success in whatever you do. It doesn't make much difference what field of work you choose, just so you do something you enjoy and do well. Satisfaction with your job and with yourself means more than a big paycheck. Of course if you are offered big money for doing well at a job you like, don't turn down the money.

There's nothing wrong with money. But don't lose your perspective about it. Money is a means to an end – something to help you have the necessities and to enjoy the comforts of life.

Use your money, don't let it use you. You don't measure your richness in life by a bank account.

Pay your debts, keep your credit good, and try to save some for the future or emergencies.

If you must gamble, never invest more in a game or on a horse than you'd be willing and able to pay for any other entertainment. Never bet your rent or grocery money on a race and don't draw to an inside straight.

I hope you're not a sissy, and I more ardently hope that you're never a roughneck. A bully is more contemptible than a coward, because the coward can't help himself.

The important thing is to be neither of those, but to be a man. That means that you neither run from nor provoke a fight. And it means that you are considerate of other people.

I don't care how you talk to your boss or your banker. They can take care of themselves and you will learn not to repeat your error. But I do care how you speak to the waitress, the maid or the clerk.

I'd rather hear that you insulted a "Big" person than to hear that you hurt a "Little" person.

Learn to be temperate and tolerant in all things.

Never let your mind or your body become contaminated. Don't smoke, drink or eat too much. Moderation is one of the most important lessons to be learned in life.

Don't worry about good manners, if you think in terms of etiquette. If you are considerate of other people and use common sense, you'll have good manners even if you aren't sure which fork to use on the salad.

To be considerate, you must respect all people as human beings, regardless of their station. This is the true test of a gentleman.

Be respectful to all people, servile to none. I hope you will never feel inferior to anyone, and so much understanding that you will never try to make anyone feel inferior to you.

Respect other people, but above all, respect yourself. Don't worry about what other people think about you, but always worry about what you think of yourself. Speak strongly for your principles, but try never to speak in anger. Only weak men need volume or violence to express themselves.

You must have a sense of humor to enjoy life, but that doesn't mean enjoying jokes only because they are dirty. It means that you will enjoy life if you see and appreciate its lighter side – and that includes the times when things are tough and you can laugh at yourself and mean it.

It is better to be neat than fashionable. Keep your nails and your neck clean and don't fret about the cut of your suit.

Be honest. You must expect some people to dislike you, but nobody should distrust you. Your word is your most valuable asset.

Don't be afraid of making mistakes, but be ready to admit and correct them. It's no disgrace to be wrong, but it is disgraceful to avoid responsibilities for a wrong. Acknowledge your achievements and your errors with equal grace.

Always be able to look at yourself in the mirror without being ashamed.

And I pray that you will be the kind of man that my father prayed his son would be. For then you'll be a man, my son.

Counting My Blessings

I am thankful to live in a country which believes in offering Thanksgiving for its blessings and thankful for the many blessings our country gives us.

I am glad I was born to the parents I had; I am thankful that my father had a long and full life and that my mother remains in good health.

I am thankful for my wife and thankful for my children and my grandchildren and my sisters, aunts, uncles, nieces, nephews and even for my in-laws, and especially for having such a nice mother-in-law.

I am thankful that we live in South Florida, but thankful too, that I have lived other places and have visited many more.

I am thankful for friends and even thankful for my few enemies for they make me appreciate my friends the more, but I am glad I am not bitter about anyone.

I am thankful to be alive, to be living where I am, working where I am, doing what I do, and to have those I have for family and friends.

I am thankful that I and mine have sufficient to eat and wear, a roof to protect us from the rain and walls to hold out the winds; lights to read by and books to read, and most of all that it is a home and that we in it not only love one another but like each other as well.

And now, if I only knew how to carve a turkey, my life on this day of thanksgiving would be complete.

Merry Christmas to All

Merry Christmas to All

Merry Christmas! This is my personal Yuletide greeting to each of you who reads this, and my hope that you will have your share of happiness on this happy day.

May all petty prejudices and mean thoughts be forgotten and good will prevail.

A merry Christmas for all of us fortunate enough to bask in the warm Florida sunshine, even in winter, and our consolations and good wishes to those unfortunate enough to have a white Christmas with sleet, slush, ice-covered streets, cold feet and frozen noses.

I wish a special Christmas to people who are lonely, especially to the aged or infirm who are in nursing or rest homes, to men and women without mates or without children, to the elderly without grandchildren and to those separated from their children.

Light the candles for the men who collect the garbage and clean the septic tanks, the mechanics on their backs beneath the car and in the grease pit below, the test pilots, the men who climb the utility poles and descend in the manholes, the nurses who change the bed pans – to all those who perform disagreeable or dangerous tasks for the welfare of others.

Pat the head of teachers who patiently try to instill knowledge in the heads of youngsters and of parents who try to instill ideals in their hearts. A blessing, too, on all youth leaders.

Sing a carol for the farmers, the oilfield workers, the musicians, the ditch-diggers, the undertakers, bakers, plumbers, librarians, the engineers and electricians, and even the bankers – to all men and women who work for a living and especially to the working mothers.

A bright Christmas card to the boy who strikes out, the student who flunks the test, the receiver who drops the pass, the player who is left on the bench, the man who is always last in line, can't pronounce the meals on the menu, and never has correct change for the phone or bus.

Light the Yule log for people with insomnia, for men with nagging wives, to women who can't help nagging, to night workers, to miners who toil in tunnels deep in the earth, to deep-sea divers and to sailors.

A joyous Noel to children who are mentally retarded or physically handicapped, and to the people who care for them or teach them; and the same to all people with handicaps of any nature, to those who are paralyzed and to those who nurse them.

Merry Christmas to those who must work on Christmas – the newspapermen, the elevator operators, the utility repairmen, the cooks, chefs and dishwashers.

I offer A Prayer for the soldiers far from home, wherever their duty may have called them.

And I offer hope that bombs and bullets will not prohibit some observance of this Christmas.

And the prayer, of course, extends to our Prisoners of War and Missing In Action, wherever they may be.

Drink a wassail to the women who scrub the floors, the maids who make the beds and clean the lavatories, the night watchmen, the firemen and police who put their lives on the line so that we may be safe.

And let us not forget the mailman who brings us good tidings and bills with equal efficiency and especially we bless those who are friendly to us.

Sing a happy song for the telephone operators, waitresses and clerks who are cheerful and polite to customers, and a pox on those who are curt or impatient.

Ring the bells for special people such as the cab, ambulance and bus drivers.

And for disc jockeys, horse trainers, bartenders, filling-station operators, shoeshiners, dentists, bookies and book-keepers, all those who toil in hospitals and particularly to those few doctors who still make house calls, even after hours; and forget not the volunteers, young and old, who help in hospitals and nursing homes.

A sprig of mistletoe to the wives who change the diapers, sweep the floors, wash the dishes, get the kids off to school, baby their husbands and cook the meals, but never demand time off or overtime pay.

Blessings on the Salvation Army, Goodwill Industries, Miami Rescue Mission, Camillus House and all others who help the needy without patronizing them.

A big Christmas bonus to all grandparents and uncles and aunts, to high school students who want to go to college, to stable hands and the man who cleans up after the elephants, the clowns in the circus, to the preachers, priests and rabbis who are concerned about their brethren in all seasons, and to the young men who are gentlemen, and the young women who are ladies, no easy task these days.

To each of you:

Merry Christmas, Happy New Year!

Larry

MOTHER, THIS IS YOUR HAPPY FAMILY IN MIAMI--
WHAT FAMILY?
WRONG NUMBER
MISTLETOE
OH, MOTHER, I FINALLY GOT YOU--YOUR HAPPY FAMILY--
WHAT'S BECOME OF OUR HAPPY CHILDREN?
DISAPPEARANCE ACT
WHERE'RE MY HAPPY CHILDREN?
I'M CARL-
I'M EVELLEN--MIRIAM'S POUTING SOMEWHERE-
WHAT'S THAT NOISE, LAWRENCE?
THAT'S HAPPY MIRIAM WISHIN' YOU A MERRY CHRISTMAS
ROBIN HALL

The Robber Got Robbed

It was a quiet Saturday evening in the Pleasure Bar, a neighborhood tavern at 2685 SW 28th Lane in Miami. Phil Osborne, the owner, had gone out for dinner and his old friend, Charlie Link, was taking care of the store for him.

There really wasn't much for Charlie to do. Earl Reardon, an electrician, was sitting at the back end of the bar and Bob Stanton, a retired Marine, near the other end, close to the front door. There was one stranger, a tall guy, having a beer.

This fellow finished his beer, stuck his right hand in a pocket of his jacket and walked over behind Earl.

"This is it!" he announced in a loud, nervous voice. He pushed his right jacket pocket in Charlie's direction. "This is a holdup!" he said.

Charlie figured maybe the guy was trying to play a joke on him, and he grinned at him.

"You want to get killed?" the man asked. "Hand me the money!"

Charlie opened the cash register and handed over about $25 in bills. Stepping forward to pick it up, his right hand still in a pocket, the man brushed an arm against Earl's leg.

"Keep your hands off me!" said Earl.

"Shut up or I'll shoot you!" snarled the holdup man.

"Okay, but keep your cotton picking hands off me!" said Earl.

"I'm sorry," said the robber.

Just then Phil walked in. He walked past the man and went behind the bar.

"There's trouble," Charlie whispered.

Phil saw the stranger standing there back of Earl. They are a little clannish in the Pleasure Bar. They are polite to strangers, but not cordial. Phil figured that this stranger had a snootfull and was giving Charlie or Earl some difficulty.

"Don't serve him," he told Charlie.

"I've already served him," said Charlie. "I served him all our cash."

"That's enough!" barked the fellow. "Everyone get in the men's room!"

Charlie add Phil and Bob and Earl stared at him. Nobody had ever ordered them to do a thing like that before. Besides, they knew the men's room wasn't big enough for all of them. Nobody moved.

"I said march into the men's room!" shouted the man, waving his pocket. Nobody moved. "You want me to shoot you?" he asked.

"You must be crazy," said Phil. "Get out of here!"

The man backed slowly to the front door, keeping the small group covered. "I'm not crazy," he said as he opened the screen door. "You guys are the crazy ones." And he headed across the street where his car was parked.

As soon as he left, Phil and Earl bolted out of the back door and around the side of the building. They thought they could at least get the license number of the getaway car and see where it headed.

Bob ran out the front door. He saw the man take his hand from his jacket pocket. He was not holding a gun and the jacket didn't sag, as it would have if there were a gun in the pocket.

"Hey you!" Bob bellowed. "Come back here!"

It was an order delivered in the tone he had used during more than a quarter of a century as a sergeant in the Marines.

And the man, who was nearly at the car, stopped in his tracks. He turned, and walked back. If he hadn't been a private at one time, he certainly reacted like one now.

"Stand against the wall!" Bob ordered, and the robber did.

Just then Phil and Earl rounded the corner.

"Give me my money!" Phil demanded.

The man reached in a pocket and pulled out some bills. Phil snatched them from him.

"Wait a minute!" the man protested, in a plaintive way. "Some of my own money is in there!"

"You get the hell out of here!" said Phil.

"Yes sir," said the man, and he ran to his car and sped away.

After the police had been there, Phil counted the money. His own receipts were there. And so were 10 other dollar bills, neatly folded and held with a paper clip. He had actually robbed the robber.

And the robber, I am sure, is still trying to figure what went wrong. He followed the rules, but things hadn't turned out right. But then, as he said, people at the Pleasure Bar are a little crazy.

TROUBLE IS STILL AFRAID TO GET NEAR CAESAR WHO DOESN'T KNOW THAT A WAGGING TAIL MEANS FRIENDLINESS

Lost in the Wilds of Queens

Somewhere in Queens – Don't worry about me. I'll be all right. I'll send up some flares and the Coast Guard will find me one of these days and I'll be back home laughing about this experience.

Right now, I'm lost in the great big borough of Queens. A few minutes ago I thought we passed the Statue of Liberty, but anybody knows that it isn't in Queens, so it may have been a statue of Queen Victoria.

The way this got started, we were in Nutley, N.J., where my older sister lives, and we were going back to Jamaica, which is a community sort of like Allapattah or Edison Center, which is on Long Island, which is like Normandy Isle, which is in Queens, which is sort of like Miami, which is in Queens County, that's kind of similar to Dade, all of which is part of New York City, and there's nothing like that in the whole world.

You can see from that why it is no simple thing to find a plae in Jamaica when you're in New Jersey.

We had no trouble getting out of New Jersey because we found the Lincoln Tunnel, that takes you right to Manhattan. That also is in New York City.

To make sure we couldn't get lost, I asked the man who collected the toll how we got to Queens.

"Yourbestbetistagota 34th St., Right?" said the man, talking as if he were spitting out machine gun bullets. "Thenyabettagoleftafewblocks, Right?"

"Left?" I asked.

"Right!" said the man. "Gothruoughthecitytillyaturntoya left, Right?"

"Right," I said.

"Thenyagothrough the QueensMidtowntunnel and yastayonaparkwaytilyawhereyawannago, Right?"

"Right!" I said enthusiastically.

We drove on a few inches. "What in the world was that man talking about?" my wife asked.

"How do I know?" I said. "He must be some kind of foreigner."

We got to the next corner. Naturally, we turned right. Right?

I don't know how we made it, but we found the Midtown Queens Tunnel. It was easy to reach Queens after that, because you can't get out of the tunnel until you're in Queens. But after that, you've got a problem, trying to find Union Turnpike which you want to take until you reach 164th St. so you'll be near the corner drugstore near your wife's sister's house.

First, we wanted to turn off the tunnel parkway at Queen's Boulevard. My wife, who was driving, stayed in the middle lane until we saw a turnoff sign, since we didn't know if we were to turn right or left. We finally saw a Queen's Boulevard sign, on the right and we spent half an hour and made 1,000 motorists mad trying to get over to the right lane.

Did you ever hear 1,000 horns squawking at you simultaneously? It's loud, right?

But it turned out that the sign had said, "Queen's Boulevard, Keep Right," and we already had turned off before we realized our mistake and we had to travel around the world, through several unexplored countries, before we could get back to the parkway, and when we got back on it we were headed for Manhattan and New Jersey.

There is nothing better than a tour of Queens by auto. It has some most fascinating scenery. If I'm ever rescued, I may write a book, "Inside Queens," or "Around the Borough In 80 Days or More."

The 'Old Man' Is Back

If you don't mind, Betsy Poller, would you remove those – uh, feminine things – from the back of my chair? I know it is common for girls to rinse things out and hang them up to dry in the bathroom, kitchen, or whatever space seems most readily available. After all, I was raised in a family with one mother and three sisters, and I currently live with one wife and two daughters, all female, so I know something about the habits of females.

But this just happens to be my desk and my chair. And just because they let you take my sapce in the paper during my late lamentable absence does not give you a lease on the property. At the risk of seeming to unchivalrous, you take your things and get back to the women's department where you belong.

The old man is back at the homestead. And if I can ever restore this desk to its normal cluttered condition, I will feel at home again.

Incidentally, far too many people tell me, Miss Poller, that you wrote some very clever material in my space. I guess that was all right for readers who go for that sort of thing, but, frankly, I get a little tired reading about somebody else's family life and children. True, I enjoyed your column before they put it in my space, but that happens to be reserved for my family life and my children. I am sure I will find you as interesting as ever when you reoccupy your own space.

Now, if I can get all those books on the sex life of octogenarians and hobbies for the wheelchair taken off my desk, I will return some of my own reference material.

This included priceless notes, books and clippings on such subjects as "Old Songs My Grandmother Sang to the Boys in the Back Room;" a pamphlet on "How to Tell a Royal Poinciana Tree From a Poke Salad and Other Helpful Hints on Flowering Trees and Tropical Plants" and a book, "Memories of Oklahoma," by the Dalton Brothers with a foreword by Pretty Boy Floyd.

You perform an invaluable service for our senior citizens, who are not all so senior, Mr. Robert Peterson. I used to read you myself, just to be ready for my 40th birthday and the rebirth you promise. (Some of my close relatives, such as my mother, think I passed that period long ago, but my mother's memory is getting bad.)

The point is that you write useful and entertaining articles, but it so happens that I have my own favorite pioneers, grandmothers and long-lived friends, and this space is our meeting ground. You take your oldsters to your own meeting house and leave me alone with my own wrinkled self.

And now, you there with The Los Angeles Times decals all over your attache case, get those big feet off my typewriter. I recognize you, Mr. Jack Smith. I've been gunning for you ever since the nurse came into my hospital room giggling all over. "That Mr. Smith in your place in the paper is the funniest man I ever read," she giggled. Then she practically rolled on the floor telling about a column you had done about riding in an airplane. I don't see anything so funny about that. Besides, I wrote a hilarious piece about riding in an airplane years ago, before you had taken your first scooter ride. The nurse gave me such an upset, I had to be put in an oxygen tent two days.

You get back to your smog and your California oranges and leave the Floridians to me. I may not write as well, but I use Florida orange juice.

I will face up to it. You folks who have substituted for me in my absence have been very good. I know it. The trouble is, my readers were finding it out. So were my bosses. This space isn't big enough for all of us, or even two of us. This is the only column I have, and I want it all to myself.

So you folks, scoot! Git! Split! Get going! Sell your words at some other stand, and I'll once again admire and love you.

But right now, get off my property!

Dogs, Cats and Other Autocrats

We've Been Adopted

We didn't adopt our new cat. He adopted us.

Maybe his original owners, Mr. and Mrs. Norton Eisenberg, had explained to him that we had just lost our pet, Caesar, and that it was going to be up to him to take the minds of three children away from grief and maybe cheer up two adults as well.

Maybe they had drilled him on the importance of joining the family and doing a good job of building morale. Or maybe he liked our looks and decided he would like to live with us.

We would like to think that it was the latter reason, although he is accomplishing the former objective.

From the second my wife held him in her hands we were his.

When she put him on the floor he made a minute inspection of the entire house, as if to say that since this was going to be his permanent home he might as well learn the layout. He gave such a close inspection, even looking under the beds and behind all drapes, we were worried about whether the place would suit him.

We half expected him to order us to remodel or rebuild or to change the furniture around.

When he had finished his tour, with my wife and me trailing behind anxiously awaiting his verdict, he returned to the girls' room, leaped on Miriam's bed, curled up and went to sleep.

He was home.

When the children came from school, Junior was in that room and on that bed – not still, but again.

The girls went in first and their squeals brought Carl right behind them.

He called me at the office.

"Guess what we found!" he shouted. "A baby Caesar!"

Oh yes, that name, Junior. Naturally it means Caesar Jr.

It isn't a very good name, but he seems to be such a very good kitten that I think he can make it respectable.

Enemies Become Pals

It seems like a long time ago that this little ball of mewing fur appeared at our house and moved in as if the place belonged to him and our dog prepared for a battle to the death over that question.

It is hard to believe that they were ever enemies, because they are such good friends now.

Trouble finally accepted the kitten, because the kitten simply refused not to be accepted. You can't stay mad at somebody who insists on liking you, and from the beginning Junior, the cat, made it plain that he liked Trouble, the dog.

The only trouble was that he teased Trouble and woke her up at an ungodly hour to play games and disturbed Trouble's naps during the days and kept her up after Trouble's normal bedtime. The dog had no choice but to wake up. At first we were afraid she might try to kill the cat, but now it is as much a game for her as it is for the kitten.

The only thing is that Trouble still gets mad when the kitten wakes her up so early, but she quits growling after a minute and the game is on.

This consists mostly of the dog lying on her back or side waiting for the cat to attack. The cat attacks in all manners of ways and from all directions. Sometimes she goes for the tail, other times she aims for the neck, sometimes she heads for the nose, often she grabs an ear. She also likes to put both paws around the dog's neck, from either the underside or the topside.

When Trouble walks into a room the cat leaps on her back just like a tiger trying to fell a water buffalo.

Between battles, they nap together, the kitten often resting his tiny head on the dachshund.

The kitten makes these battles seem very realistic. One of her tactics is to suddenly arch her back and stiffen her legs, then jump straight in the air, but sideways.

I know this sounds as ridiculous as saying the Wright brothers' plane could really fly, but that kitten can jump sideways in that funny position. After two or three leaps he is back on the dog. The dog has been watching this exhibition with open-mouthed amazement and is caught completely unprepared.

This may sound as if Junior is as smart as a tiger, but the truth is that he is the most awkward cat that ever lived. He has no sense of distance or depth.

On the hassock, the divan, a bed, a card table, he invariably will misjudge a leap sooner or later and make an inglorious drop to the floor. What's more, he can't even land on his feet. Most of the time he lands on his back.

From the goofy way he acts much of the time, I suspect he landed on his head at a very early age.

Another thing. He is talking more. And louder. His voice isn't as deep as Caesar's yowl used to be, but he yammers more. Maybe he's going through his teens which, I hear, are very talkative in all species. I hope so, because I don't think he really says anything.

The Temporary Kitten

When our daughter brought home a kitten, saying she was going to find a home for her and was merely offering a temporary pad, my objections were loud and clear. We already had one cat, and I made it plain that he was plenty.

The kitten looked like a freak at first sight. When in motion, which was all of the time she was awake, she was a streak of black and white. But at close quarters you could see that the black and white, while predominant, was completely irregular and that it was the background for countless spots and streaks of brown and orange. Some of them looked as if a painter had cleaned his brush on her; others as if somebody had flipped paint at her from a distance.

Despite this, she was cute and she was funny. She regarded Caesar as her best friend and playmate, although Caesar regarded her strictly as a pest.

Her favorite game was to play war with Caesar. Sometimes she would stand on her hind feet and box Caesar until the older cat growled or struck back, at which she would beat a hasty retreat and resort to guerrilla warfare. She was an expert at hit-and-run tactics, bothering the mighty Caesar as effectively as a fly pestering a bull.

She would dash out from under a chair, nip Caesar's tail, then scoot in the opposite direction, crawl under the divan and dart out to take a swipe at Caesar's head as she went by. She would circle the room, thinking she was as quiet as an Indian but actually as conspicuous as a black and white rubber ball.

Caesar knew where she was at all times, but he pretended not to care. He even tried to fake sleep while these attacks were made, but sometimes he lost his temper. Then he would try to catch the kitten when she came within striking distance, holding her under one paw while he nipped her on the neck or stomach. But Caesar never bit hard and the kitten squirmed like a fish, meantime lashing out with the tiny but sharp claws on all four feet until she escaped or Caesar let her go. Caesar then would heave a sigh of relief, thinking he had taught the little nuisance a lesson. But the kitten would dash for cover and regroup herself. This time she might come out with arched back, standing sideways rather than facing Caesar, and approach him by short hops on stiff legs. This was the most fantastic military maneuver I ever witnessed and I still can't figure its purpose unless she thought it made her look like a fearsome monster.

Whatever the tactics, she would return at once to the battle.

Her only weakness in this type of hit-and-run fighting was her timing. Sometimes she would charge at Caesar and make a leap, intending to land on his back, and go clear over him or land in front of him. Other times she would get in a chair and leap at Caesar as he passed, only to miss him completely and sprawl on the floor.

There never was a dull moment when that kitten was awake. She even took over the chaise lounge, the longest and largest seat in the house, as her own private chair.

Meantime, I kept asking Evellen when she was going to give the kitten to that friend. Meantime, too, when I was alone, I played with the kitten and was flattered when she occasionally leaped on my lap to take a nap. Her purr sounded like a miniature buzz saw.

Then came the morning when I was dressing and the kitten didn't come in to help me put on my shoes. She thought it was her job to assist with the laces. Although I pretended to be upset by the way she did this, on this morning I finally went to the door and called my wife, asking where the blankety-blank kitten was.

She told me I would be delighted to know that the kitten had a new home.

Evellen still can't understand why I bawled her out for giving our kitten away.

Funny thing, too. Even Caesar keeps looking around for his tormentor. If I didn't know better, I'd swear that Caesar missed her.

Moral: If you don't want a kitten, don't let one get near you.

The Duchess Is a Tramp

I wish that Eleanor Hart or Dorothy Ricker or Amy Vanderbilt or some other authority would give advice to young dog ladies who are straying from the straight and narrow path.

It grieves me to report this, but I am afraid that Duchess, my dachshund, may have wandered into the primroses.

She comes from a good family and goodness knows we have tried to give her a good home since she came to live with us.

Her behavior at times has been slightly giddy, but the sort of thing you expect of a growing girl, an adolescent who can rock 'n roll and considers adults immature.

We haven't been strict with her, letting her roam the neighborhood and pretty much keep her own hours just so she came home for meals and for the night.

A Wide Circle of Friends

As a result, every kid within three blocks of our house knows Duchess, and I am known only as the man who owns her.

But she never abused this freedom, and I felt that my confidence in her was justified.

Every evening when I came home from work, Duchess was standing in the front yard, hardly able to wait for the car to stop in the driveway so she could greet me with wet tongue and loud barks.

Every night at dinner, Duchess would make the rounds at the table, and she taught both the twins and me terrible habits so that we kept slipping morsels of food to her until The Woman of the House caught us at it.

After that, she would help put the babies to bed, then would station herself on my lap, to guard me until I went to bed. She would then go to sleep on the front room divan.

That's When She Deviated

That's the way it has been, but the other night Duchess didn't follow the pattern.

My wife met me downtown and we went to Crandon Park with some friends to cook steaks and it was late when we got home.

I missed having Duchess in front to greet me, but I figured she was inside, sleeping like a good girl.

But she wasn't inside and the young woman who was looking after the babies said she hadn't been in the house for several hours.

Looking back, I suppose that Duchess had come home at dusk, had waited patiently for me to come home, and after she realized I wasn't going to appear, maybe went out to drown her sorrows. I don't know, of course, but I feel that I may have been to blame for her downfall.

But regardless of the reason, Duchess wasn't home. And she didn't come home all night. I went out several times to call her, but got no response, and I got pretty worried about her.

Early the next morning she was at the front door, begging to get in, and she was the most worn-out looking dog I ever saw.

I should have given her a lecture, then and there, but she looked so pitiful that I didn't have the heart. She did have the decency to look ashamed as she crept in and barely managed to leap on the divan.

The moment she landed, she went to sleep. As soon as she was asleep she turned on her back, poked her legs up in the air, and began to snore, a most shameless hussy.

She was a disgusting sight.

I had expected her to at least offer some explanation, but when she went to sleep just as if she had done nothing wrong, it hurt me. I went over, meaning to smell her breath, for I strongly suspected that she had been in some dive where young ladies shouldn't be.

Duchess heard me approaching and she opened one eye and gave me an expression which only adolescent girls can give a parent. "All right, what are you going to accuse me of?" her expression said. "I suppose you're going to cross examine me now, as if I were a little girl. All the other girls my age stay out and their parents don't heckle them all the time. Good grief, you'd think I was a baby. If you don't trust me, I'm going to start doing the things you think I do ----."

I say I could see from her expression what she thought, and I left her alone. She never has told me where she was or what she did.

But I'm very afraid that Duchess is no lady.

Sad Tale of the Pup

The next time I read a story about a family leaving one of its children in a filling station or accidentally deserting an offspring, I will not snort and say that such a thing could not happen to normal parents who love their children. I will be sympathetic and understanding.

You see, we deserted one of our pups the other night.

Now a pup isn't on quite the same level as a child, at least not in my family, but you still don't go around abandoning them. Anyway, we don't, or anyway we never thought we would.

On this evening, we went to Coconut Grove Bayfront Park, as we do frequently, for a picnic dinner and to enjoy those cool breezes that come whipping in from the bay even when the air everywhere else is still and sultry. We took the pups along and let them chase each other and sniff along the edge of the water for crabs or whatever it is that pups sniff for on the edge of the water.

When we were ready to go home and had assembled children, pups, picnic equipment and assorted odds and ends, my good wife noticed that we had left a box at the table. It was filled with things like mayonnaise and pickle jars, knives and forks and odds and ends like that.

Saying something to the effect that nothing ever was done right unless she did it personally, she took one of the girls and went after the box. At least one of the pups had been in the car. At least one of the pups went with my wife. Some of the kids stayed at the car. Some went with my wife. Some wandered off. Things get sort of confused when everybody is trying to get organized at such times.

Anyway, we got the box and loaded everything in the car and reassembled the kids and got them in the car and went home and when we got there, Caesar was sitting in the middle of the sidewalk in front of the house, waiting and growling, and the kids erupted out of the car to go grab Caesar and somebody else turned on a light in the house, and I was yelling at everybody to come back and help carry things and my wife was yelling at everybody to come help carry things and we loaded people up as we got them back at the car and we finally got everything in the house, including Caesar.

There had been a lot of dogs mixed in with this confusion, but when we finally got the horde and the odds and ends in the house we took no census, and I will explain why.

For one thing, we were busy yelling at the kids to get in their pajamas and go to bed and I think maybe my wife was trying to get them to take a bath at the same time, and for another thing, along about bedtime our pups disappear.

This is because they don't want to be put in their quarters out in the Florida room.

Sometimes Cheeky will run in our bedroom and get on the bed, pretending to be sound asleep when we find her, in the hope that we will let her stay, and sometimes this works. Trouble is smarter, for she knows that often we move Cheeky from the bed to the dog quarters, so Trouble will get under the bed and not make a sound until all lights are out and everyone asleep, at which she will jump on the bed as quietly as possible and hope she can stay for the night.

At other times, one or both of them will hide under the divan or behind a chair or go in the girls' room or the boy's room. The idea is that they make themselves scarce at bedtime, and we don't think anything about the fact they aren't both together in the most prominent place in the house, as they usually are.

And so it was that we didn't actually miss Cheeky this particular night for several hours. I had gone to bed first and when my wife came in, Trouble and I were sharing the bed. My wife woke me up by calling for Cheeky, who usually was first on the bed, then I got up and we looked in all the hiding places, then we woke up the kids to ask if they were sure we had brought Cheeky home with us, and everyone said we thought we had but none of us was sure. And at last the awful truth had to be admitted, that we had abandoned our pup.

We went tearing back to the park, but of course the pup wasn't there. We returned home, full of remorse and worry about what had happened to the little thing and what she must have thought when she found we had left her. Most of all we hoped that she had not wandered into the streets and been run over by a car and most of all we hoped that she had been seen and taken in by people who would be kind to her.

Trouble Remembered

Little Trouble is dead.

Trouble was a bronze-red miniature Dachshund presented to us by Mrs. Alice Lockey six years ago. She was a member of our family.

She was funny, lazy, silly, and slobbery, but she was friendly, devoted, beautiful and lovable.

She liked to sleep under a cover even if the weather was only slightly cool; she could stand up and even waltz around on her hind feet when begging, but pretended to be dumb if you tried to teach her to sit at command.

She would sleep on the divan, staying in her favorite cover – an old bedspread – after the rest of the family was in bed when I stayed up to read or watch TV. Whenever I glanced her way, her tail would thump like an Indian tom-tom.

Sometimes she would get impatient and try to talk to me, even tugging at my pants leg, to tell me it was time for bed. I would put her cover at the foot of the bed and she would burrow under it. Then she would sleep against my legs.

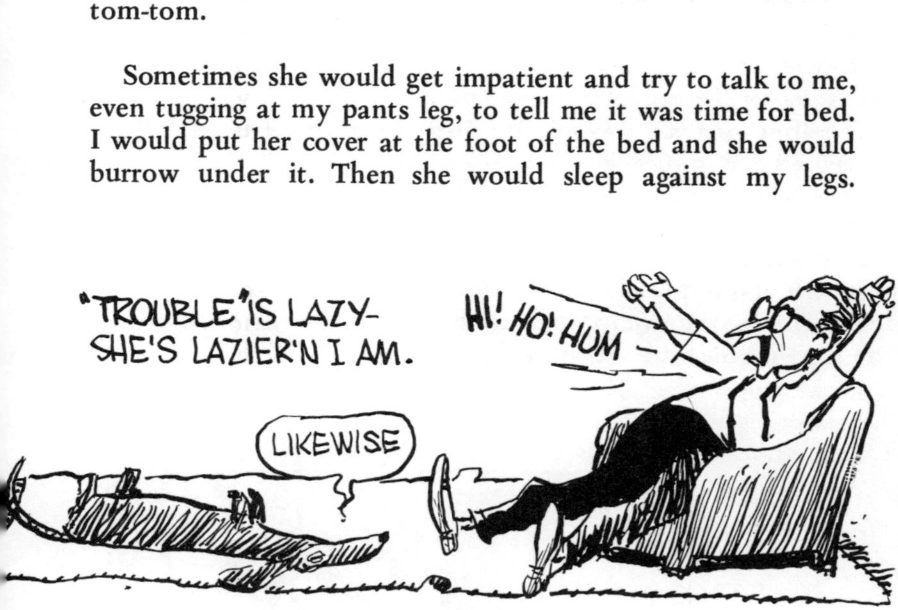

Some cold nights she would wait until she was sure I was asleep, then sneak under our covers. She always looked surprised when I uncovered her in the morning; surprised that I had found her and pretending surprise to find where she was.

Once when I was in the hospital for a month, she wouldn't go into my bedroom at night. She slept on the divan until I came home. On my return home, she acted as she did each year when the family returned from vacation. On those reunions, she would wag her tail, shake her hips like a hula dancer and go crazy trying to jump on everyone and lick all the faces and hands within reach.

She was a coward. She was ferocious only in her sleep, when we could tell from her growls that she was fearlessly facing some terrible monster.

She would bark at other dogs, the garbage men, the trash collectors, passersby when she was safely inside the house or if they were at a sufficient distance, but in actual confrontation, her only weapon was a wagging tail to show she had been kidding.

But she had courage.

When we took her to the veterinarian, as we did twice in the past three weeks, she would be frightened, but make no attempt to escape or struggle; on the operating table when he examined her, she would tremble in fear, keeping her eyes on me or whichever member of the family was there, to make sure this was required of her, and accepting her fate; accepting whatever the doctor did to her, even when he took out the long needle and she knew its purpose.

ONCE INSIDE, TROUBLE IS ALL FRIENDSHIP, LEAPING ON ANY VISITORS LAP AND TRIES TO LICK THEM IN THE FACE.

She had a courage in her small, trembling body greater than that of a large dog trained for attack or combat.

The doctor was puzzled about her ailment. After shots and with pills she improved, then relapsed; then repeated the cycle. She improved remarkably the second time and even played with the cat. But she became listless again Saturday day night. and refused to eat. When company visited Sunday afternoon, she didn't even leave her divan corner to greet them, an unheard of thing since she was our family greeter and hospitality chairman.

Sunday night the girls were going to take her for a walk. Outside the door, she sat down, and when they urged her on, looked at them with pleading eyes. They brought her back in and put her at the foot of my – of her – bed.

An hour later, when my wife looked in, she was dead.

She was wrapped in her own cover when we took her to the veterinarian Monday morning.

He found that she had died of cancer of the intestine. He doubted that she had been in pain.

This morning there was no lump under a cover at the foot of my bed. Usually when I get up, a thumping noise from under that cover always signaled that Trouble had received the message and would join me as soon as she could untangle herself.

Sometimes she fell off the bed trying to get uncovered, and when I laughed, she wagged her tail and was pleased she had amused me.

She was a silly dog.

Vacations and Stuff

Bliss, Then Calamity

Caruso's Resort, Ontario – Although this was one of the most enjoyable stops of our entire vacation, we will remember it in terms of our worst experiences. It was here our car stalled and it was here I lost our money.

It is near an Indian reservation, at Espanola, about half-way between North Bay and Sault Saint Marie. We arrived early enough on a Saturday afternoon to enjoy several hours by daylight.

We had a roaring campfire, and I was lucky enough to get our lantern to work, so it was real cozy that night.

We liked the place so well we decided to stay an extra day and Sunday was just about perfect.

My wife spent much of the day taking a sun bath, her first real relaxation so far. The children spent much of their time swimming and fishing.

By mid-afternoon, Carl had convinced himself that all the fish were in the middle of the lake and he talked me into letting him rent a row boat. I made him pay half of the rental, which I believe was $1 an hour.

These negotiations awakened my wife, who does not believe children should be in boats unless accompanied by parents. She finally agreed that since our children are better swimmers than either of us, that it might be all right if we watched them. We both are having trouble realizing that 13-year-olds no longer are children.

Carl and the girls got in the boat and went out about 100 yards. They had to wear life jackets, and they promised to stay near the boat docks.

After the first two strokes, Carl realized he had been booby-trapped, since he had to do all the rowing. It made him furious that the girls insisted on fishing while he was rowing.

After one hour they came in. Carl was still sure he could have caught a fish if the girls hadn't made so much noise. They said he had made most of the noise, yelling at them. Sometimes he would yell to his mother and me.

"Look!" he shouted once. "They have their lines all tangled!"

He reached for one line to straighten it out, then shouted again: "Hey! We caught a fish!"

But the fish got away.

I spent much of that day by myself. I climbed to the shore of the lake just below our camp, and sat on a big rock, and I looked at the lake and the other shore and the little islands and sometimes a motorboat or a sail boat, and the sun shining on the water and making it look like aluminum with an occasional breeze turning it into riffles. And I sat and was most content. Times like this make camping worthwhile.

The perfect weather ended during the night and it began to rain. Next morning, during a lull, we decided to pack up and hit the road, since the rain seemed to have settled in for a long stay.

We had just put the camper trailer down and hitched it to the car when the rain returned. We turned the key and the car just sat there. The battery was as dead as the well-known door nail. Some day I will find why a door nail is so dead.

This was a terrible predicament. The closest town, Espanola, was about 15 miles away. It was a small town. Furthermore, this was the first Monday in August, proclaimed a national holiday by the Canadian Parliament, so chances for finding a garage to send help were almost non-existent.

I must cut this story, but a government employee and his family were living in one of the cabins at the resort. He was foreman of a crew that had been surveying the Indian land. He had a battery charging machine – something you are as apt to find at a summer lake resort in the Canadian boondocks as an original self portrait by Moses. He charged our battery and we finally left Caruso's Resort.

Our Trek Is Checked

It was raining hard on the Trans-Canada Highway and it was slow driving. As we approached Blind River, about half way to Sault Ste. Marie, we finally found clear skies. It was 4 p.m. but we decided the time had come for our postponed lunch. It was then I discovered I had mislaid my travelers' checks.

I had cashed one of these when we checked in at Caruso's Resort near Espanola on Saturday afternoon. On Sunday, I had carried the check pad in a pocket of my sports shirt.

EARLY MORNING PILGRIMAGE

One of the beauties of camping is that it is cheap, and I had no reason for cashing another check until this Monday afternoon.

I wasn't worried at first. I knew the checks were in the sports shirt. So we drove to a roadside park just outside town and opened the tent trailer.

Then my good wife proceeded to unload the trailer. Seeking the dirty shirt, she had to take everything out.

She found the dirty shirt, but there were no travelers' checks in it. Then began the grim task of searching through every bit of luggage.

We had airline bags and little suitcases, wicker baskets and big suitcases, overnight bags and laundry bags piled on picnic tables, on the car and on the ground. We had shoes under the car and groceries under the trees.

When we had been trhough everything, including many things we had forgotten we had along, we still hadn't found the checks.

I tried to carry it off as a joke. "Some day," I chirped, "we'll all look back on this and laugh." It was obvious that this particular day wasn't the day.

"Did Daddy really lose all our money?" Miriam asked.

"He'd lose his head if he could," my wife said, which was an unkind thing for a wife to say, and not very funny anyway.

"He'd better not say anything the next time I lose something," Carl said.

At 4:40 p.m. we turned around and started back toward Espanola.

We drove through the gates of Caruso's Resort at 6 p.m.

Everyone had treated us like heroes when we left, but the welcome didn't seem so warm. It seemed to me that everyone eyed us with suspicion, wondering what kind of hard luck story we would use this time.

Their suspicions were confirmed when I told the sad tale about losing the checks, but a couple of women actually joined us at our old campsite to help us hunt.

On our last night we had had a beautiful camp fire on top of a huge flat boulder. When everyone else went to bed, I stayed up to watch it, an old weakness of mine. Eventually I put it out, by separating the burning logs with a green stick. I had leaned over many times to move a log or to scatter the coals with my stick.

It was logical that the checks had fallen from the shirt pocket and would be near the ashes.

The ashes were there, but not the checks. It was logical that they could have fallen out, but it also was logical that they burned up.

We searched the boat docks, where I had kneeled over in helping launch and tie up the boat when the kids had it, but if the checks fell out there, the fish already had digested them.

It was dark when we reached Sault Ste. Marie, Ontario, and drove over the bridge into Sault Ste. Marie, Mich. I borrowed the money from my wife for the toll.

A Modern Love Story

This really happened.

A young Miami man, born in a poor family, had worked long and hard to educate himself and to get ahead in the world. He felt sure that his big opportunity had come when he was accepted for a job with an oil company in South America, paying $600 a month.

In giving him the good news, the company representative asked if he was married, and he said no. He didn't add that he had never had the time or money to go out with girls, and actually had never had a date.

"If you happened to be married to a stenographer," said the company man, "we could have hired her, too, at $400 a month."

"Keep the job open," the young man asked.

He rushed to his best friend.

"Do you know any stenographers who want to get married?" he asked.

The friend said that offhand he couldn't think of anyone like that.

"Think, man, think! There has to be some girl who would like a nice husband and a nice job."

"Well," said the friend, "I don't think that funny looking stenographer in Joe Blow's office is married. I doubt if anyone ever asked her."

"Get her on the phone and introduce me," said our hero.

The friend did this and handed over the phone, expecting the young man to ask for a date.

"I'll make it short," he said, "I have this job in South America for $600 a month. If I had a wife who is a stenographer, she could have a job for $400 a month. That would make $1,000 a month. A smart man and wife could save a lot of money on that. Are you interested? I have to know right away."

The flabbergasted friend was even more flabbergasted when he heard what came next.

"Fine! Fine! I'll meet you in a half an hour at the Zilch Clinic for our blood tests."

It was the friend who told me the story.

"I hear from Sam a couple of times a year," he said. "He and his wife have been in South America three years now and intend to stay on several more. They have two children and are very happy together. It was an ideal marriage."

"That was a real case of love at first sight," I observed.

"I suppose so," said the friend, "but I'm not sure. Sam never has mentioned whether they fell in love."

For Scenic Beauty, Nothing Can Measure Up To Gas Pump

FOR SCENIC BEAUTY, NOTHING CAN MEASURE UP TO GAS PUMP

Grand Canyon, Ariz. – The title of this essay really should be "Grand Canyon Revisited," which would put it in the same class with some famous poems I used to study in high school literature. Those poets always were revisiting someplace.

This visit was different in many ways from my first meeting with the canyon 29 years ago. Then I was young, carefree and foolish. Now I am aging, debt-ridden and foolish. That time I was alone and on foot. This time I am with a wife and three small children.

Our introduction to this glory of Nature wasn't exactly auspicious. One of the ways we can afford this cross country expedition is by using credit cards for gas and oil. When the bills begin to come in, I'll find I couldn't afford the trip, but for the time being it seems as if we're getting the fuel without cost.

We had only slightly more than a quarter of tank of gas when we left our camp near Flagstaff and headed for Cameron, some 50 miles to the north, looking all the time for a Chevron station, that being the brand of gas which we could get on credit.

We hadn't seen one when it was time to turn west, toward the entrance of the Grand Canyon National Park, but we weren't particularly worried.

We began to realize what we were up against when we approached the East Rim Drive. As we climbed toward the clouds the marker on the gas indicator got closer and closer to Empty and the road began to twist and turn and keep on going up, and we began to get a desperate look as we searched for gas stations. By now we didn't care whether it sold Chevron on credit or boiled whale blubber for cash, just so we could get something in the tank to keep us going.

By the time we were way up in the sky, it was obvious that we weren't going to find any gas stations, unless they were built on the sides of steep precipices that dropped off on one side or on the sides of steep cliffs on the other side. Even if they had been built along the road, nothing but a mountain goat could have reached them, and anybody knows a mountain goat doesn't need gas.

The best way to approach the Grand Canyon is from the east, taking the East Rim ride as we did – if you have plenty of gas. We couldn't enjoy the scenery because we were staring at the gas indicator. The occasional glimpses I did take when we could see the majestic canyon unfolding were an inspiration. But I was thinking more of the long hike ahead of me if we ran out of gas. And my wife was thinking how uncomfortable it would be to remain parked on a narrow road, in the middle of a U turn, while I made the expedition for petrol.

We finally got to Desert View, where there is a ranger station, a lodge, cafe, refreshment and curio stores, as well as viewing platforms and a tower from which you can see most of the world.

But the ranger told us that the filling station here was closed and we'd have to drive another 25 miles to the South Rim for gas.

"Our tank is empty!" my wife cried out, great big sobs and tears in her voice. "We can't go one more mile!"

Her anguished cry touched the heart of the ranger. He told us where to find the lodge manager, who told us where to find the man who had the keys, and he opened up the station and poured life fluid into our parched automobile.

The sight of that gas hose being put into our tank was the most beautiful sight of our entire vacation trip.

Naturally the children had to go to the top of the tower. I get the jitters when I am more than 10 feet high, and this was not an enjoyable experience for me. I held on tight to the bannister as I went up the stairs and when I looked through the glass windows I stood back a couple of yards, just to be on the safe side.

On the ground level, which was the lip of the canyon, I suffered half a dozen heart attacks as our wild children dashed to the very edge. Actually, there was a gradual slope and they could safely have walked down many yards from there, but it looked to me as if they were about to fall off into a bottomless chasm.

I felt better when we resumed our ride. This time we enjoyed the views of the canyon as we rode along. At the South Rim headquarters, which has turned into a bustling village since I was here long ago, I had more heart attacks as the youngsters peeked over the edge. I was relieved to hear other parents screaming to their children not to get so close to the rim or to quit climbing on the stone walls. I guess all of us become scaredy-cats in our older years.

When I looked at this deep, wide canyon I could hardly believe that in my young and foolhardy days I had stood at this same spot, seen this same view, and had casually started down the mule trail, on a two-day hike to the other side.

I kept saying in a loud voice that I had walked to the other side, but very few people paid any attention. Seeing how I was shrinking away from the edge, it's just possible that they didn't believe me. I had a hard time believing myself.

En Route to Venice

Venice, Italy – When Adolf Hitler was an Austrian house painter, after he had painted enough houses so he could afford it, he liked to go to Berchtesgaden, a colorful village in the Bavarian Alps, and envy the more successful painter who could rent rooms with hot water and better view. Later, as a corporal in the German Army, he returned and rented a room with boiling water and a good view, but he envied the generals who occupied villas and expensive hotel suites way up in the mountains.

Then when Hitler became the No. 1 big shot, he came back to Berchtesgaden, only this time he came strictly first class. He put a high fence around his favorite peak and the surrounding mountains and the village of Berchtesgaden. It must have been a longer fence than the wall James Deering had around Vizcaya in its hey day.

Ol' Hitler made that entire part of the Alps his private playground. On top of his favorite peak, he built a fabulous layout he called Eagle's Nest. And he had houses built for his friends like Goehring, Goebbels and other top Nazis. They had utilities installed and a private railroad. Hitler even got himself a girl, and they had movies up there in the Eagle's Nest on top of the mountain.

I am speaking facetiously, but I was really impressed by the view Hitler tried to reserve for himself and his cronies. The view is as scenic now as it was then. Fifty steps below the surface are the bunkers – intact, clammy and cold – possibly the most impregnable bomb shelter in the world and certainly the most useless.

We had been given an Austrian guide in Salzburg who spoke excellent English as he apologetically pointed out Hitler's mountain. He took greater pride in pointing out Mozart's birthplace and other sights in Salzburg, but most particularly the mountains and buildings nearby where "Sound of Music" were photographed. The setting was as beautiful as the music the movie brought to it.

We had one of our few cold spells in Salzburg, with the temperature dipping to 38. For Floridians, that is zero. At breakfast, the Hotel Pitter's restaurant served the usual hard rolls and bad coffee, but this time they had orange juice on the menu and I ordered a glass for 22 schillings. I don't know how much that is, but the night before when I offered a 100-schilling bill for an airmail stamp it wasn't enough.

We crossed the Alps en route to Italy by way of the Brenner Pass, which has carried the soldiers of centuries to wars. The ride through Tyrol and the Dolomite Mountains with snow heavy on their peaks was as scenic as it had been billed, but although the Brenner Pass is rich in history, it is not in the same class as passes in our Rockies for altitude or scenery.

Tryol is mountain and meadow country, and the trees on the slopes and along the river were in full color. The river escorted us to Innsbruck, Austria, where we suddenly met the big boys of the Tyrolean Alps, waiting just outside the town for us. This is where the winter Olympics of 1964 were held, but they must have put up some extra hotels or maybe the visitors commuted to Rome, Venice and other big cities. Not even a yodel broke the quiet when we were there.

We had lunch in a lovely chalet. The river goes through the town. It is picturesque place. People are friendly. Prices for food and room are reasonable. It's too bad it's such an out-of-the-way location.

Like Hannibal and other conquerors we had crossed the Alps into Italy, and like many arrivals before us, we headed for Venice. As our bus brought us into this fabled and fabulous city, we were greeted by large "Hertz" signs, which struck me as rather strange. I would have been ready for "Chris Craft" since the streets of Venice are waterways. At least that's what I've been told.

It was a relief to find that my teachers hadn't lied to me. Venice, Italy, isn't as clean as "the Venice of Florida" (the title claimed by Fort Lauderdale), but its streets are wetter. Our bus couldn't navigate the canal streets, so we left it, but not far from the center of the city. Most folks walked to the hotel. There is a pedestrian promenade in front of businesses, but with my emphysema I was more glad to take a water taxi.

This is the first time I ever stepped out of a boat at a hotel, but it also was the first time I ever rode a cab on water.

We stayed in the Grand Principe. It was a grand old hotel with marble floors and marble tops on the tables, high ceilings, a bathtub on a foot-high marble base, and ceiling-to-floor French doors opening to balconies above the Grand Canal.

In a way it was like having a room on Biscayne Boulevard, since the Grand is the main canal thoroughfare, but I loved the sight of the boats – private boats of all sizes – yachts, sails, small motorboats, water taxis and an occasional gondola.

We didn't see many gondolas in use, and I was told that even the gullible American tourist had tired of the prices asked. Of course Venice is supposed to be a "bargaining" town, but most Americans don't like to argue about prices. Besides, my experience bargaining for a mahogany head in Cologne showed me I wasn't good at it. I paid 40 marks for something worth nothing, and thought I got a bargain.

They say that Venice's canals are little more than open sewers, sort of like the Miami River, and that may be so. They also say Venice is run down, a pitiful shadow of a once-great city-state that helped rule the world. The seat of art. The home of culture of all kinds, the center of commerce, religion, power – now all gone.

Maybe so. But looking through the windows at the Grand Hotel, I thought Venice in the moonlight was beautiful. The old town looked romantic to an old man.

A 'Find' In France

St. Maxamin, France – You won't find this town in the guidebooks or on the maps, but it was the most delightful place we visited in Europe. It is on the Rhone River south of Avignon, and we found it because it was on the highway our bus was taking between Cannes and Lyon.

From Cannes, on the French Riviera, we took a divided highway through mountains and red-leafed vineyards, past rock peaks such as we have in the Big Bend country in Texas or in John Wayne western movies, to LeLuc, a quaint village with little chimneys nosing their way through red tile roofs.

Outside LeLuc we took our first country road, a two-lane-cobblestone road that wandered through the hills, with trees on each side and beyond the trees the omnipresent grapevines. This was rural, rustic France, much different from the Paris for which we were headed.

And then we came to an area where the grapes were ripe and entire families were in the vineyards relieving the vines of their luscious burdens.

Trucks, loaded with grapes, slowly drove up the road, and at Brignoles we saw one of the wine presses to which they were going. People drink a lot of wine in France, but there's a lot of wine being produced. Somehow, the grape pickers seem to enjoy their work more than cotton pickers I have seen.

And then, in the middle of this ripe grape country, we found St. Maxamin.

We undoubtedly would have passed through it, regarding it only as another of those unnamed quaint villages, if it hadn't been for a sign at the edge of town: "English Spoken Here."

Our bus came to an almost automatic stop when it saw that sign. After Jacques, the driver, opened the bus doors, we flocked out as happily as if we had reached Paris. But if this rustic village impressed us, we didn't impress the village much. There was no band to greet us, no mayor to present us with the key to the town hall. The people went about their business as if we hadn't arrived.

There was a town square in St. Maxamin. There were tables and chairs under trees and shops around the square and the narrow streets that went out from it.

There were quite a few people on the streets, though you'd never call them crowds, and we soon learned that it was market day and the farmers had brought their produce in to sell. There also was a flea market or bazaar in full swing at one end of the main street, where one of the local citizens was trying to auction off watches and sundry other goods when we paused to listen. None of us trusted our luck with a French-speaking auctioneer, however, and we left without bidding.

My good wife and I stopped in one of the shops and had the lady in charge cut us a generous slice of cheese and several pieces of lunch meat that must have been a spiced salami. And I bought a big cluster of purple grapes and two firm red apples.

We took our purchases to one of the outdoor tables in the square. After adding a long loaf of French bread we ate our cheese, salami and bread washed down with a bottle of wine from a shop on the square.

These people were so friendly that it was a joy to do business with them – but it was an even greater joy to eat what they sold. The grapes were the size of golf balls and spouted juice when you bit into them.

If I like outdoor cafes on old town squares, and I do, I had a pleasant surprise up the road a piece at Avignon, famous as the place where the Popes lived when they moved out of Rome for 70 years. They had an enormous town square, on which were tables for a dozen sidewalk cafes. As usual, I am writing too long, so I will leave Avignon until tomorrow's piece, which will be the next to last of this travel series, proving that all things do come to an end if you wait long enough.

Camping Days Over

This is the time of year when I usually begin to look at the travel ads in the magazines, scrutinize the travel section in the Sunday Herald, and send off notes to various states for literature about their attractions and camping facilities.

Last week I received four review copies of the Rand McNally vacation publications. These include one on campgrounds, another on travel-trailer grounds, a guide to western campgrounds and trailer parks, another, "Vacation Guide," on the attractions in various areas of the country.

Those books aroused my camping blood, just as the seed catalogues in winter make a gardener impatient for spring.

But no more.

Like a dieter passing a candy counter or an alcoholic going past a bar, I try to avert my eyes from temptation and keep thoughts of forbidden pleasures out of mind.

The days of family camping have ended.

It has been a long and enjoyable family tradition. My wife and I started it before the children were born, with a hectic trip through New England during which we camped out only one night, but that was the beginning. We camped when the children were toddlers and found that they were entertained and better behaved on a camping trip than at home. From our friends we accepted (but did not deserve) praise for our bravery in making such hazardous journeys and enduring such hardships with such small children. We knew, as our kind friends did not, that the hazards and hardships were nine-tenths myth and that the youngsters thought the rugged gypsy life was more fun than a trip to the zoo.

We've covered every state except Alaska and Hawaii and that shouldn't count because they weren't states when we began these excursions. And if we missed two states, we visited two other countries – Canada and Mexico. We got real neighborly with our neighbor to the north and have been in Nova Scotia, New Brunswick, Ontario and Quebec. We would have gone to Prince Edward Island, but we missed the ferry.

I have known all the time that these family idylls could not go on forever. I was afraid we might not be able to hold the group together last year, for it was noticeable that the girls didn't enter into the pre-vacation discussions with their usual animation. I have a suspicion that if we had taken a vote by secret ballot, the vote would have been three to two against, with only Carl and me voting for it.

Still, everyone enjoyed the trip once we were under way. We had more rain and mud and other difficulties than usual, but I think we all look back on the 1969 trip with fond memories.

But for the 1970 summer, Carl, the 16-year-old boy, has a job at a lodge in the foothills of the Smokies, near Maggie Valley in North Carolina, Miriam, the 14-year-old, has been invited to visit an aunt in New York and a teacher in Ohio. The other daughter wants to be with her own friends.

And that leaves the old folks at home – and my good wife is talking about visiting relatives.

I have a sneaky suspicion that this gypsy may end up a hermit, alone in his own backyard this summer.

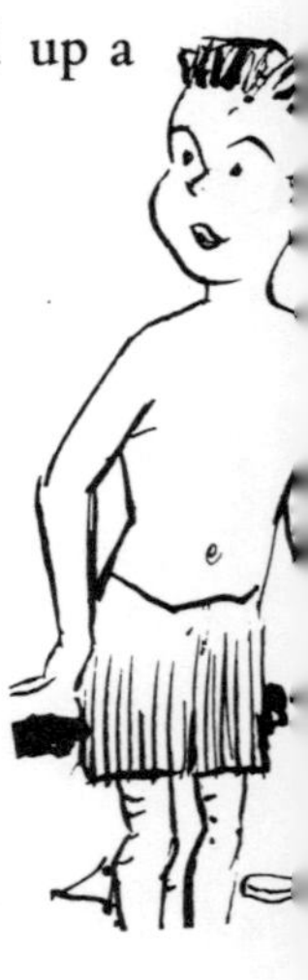

The Thompsons at Home

Gate Sags, Doesn't Swing

All I had to do was put the gate between two posts.

First I hung the gate with a couple of small hinges. The next morning my gate was flat on the ground. One of our twins had made the mistake of trying to swing on it and the gate collapsed.

Then I got a couple of big hinges. They were big enough and strong enough to support a barn door. They were about six inches long on each side.

Of course since the post was only a four-by-four that left a lot of hinge left over. It doesn't look very good, but it holds up the gate.

My real trouble started when I put the latch on the other end of the gate. I had one of those kind with a bolt that is supposed to slip into an opening.

I screwed the two parts on very good. The only trouble was that my posts were in crooked and the bolt didn't reach the opening it was supposed to go in.

As a result, I have a swinging gate. It really sags more than it swings. The bottom touches the ground when the gate is opened about a foot, and you have to edge through sideways.

You need a handyman to do any odd jobs around the house? Well, don't call me.

Now He's Been Told Off

I have been lectured by experts in my life. As a boy, my father and mother, my older sister, most of my teachers, a few principals and the scoutmaster all had turns at me.

In college, I was lectured by professors, numerous co-eds, and house mothers, fraternity brothers, and irate readers of the college newspaper.

As a working man I have been lectured by the only species that is tougher than a top sergeant, and naturally I mean that ogre called a city editor.

I've been lectured by some of the toughest in the business. And, of course, at times I have been spoken sharply to by various creditors, irate readers of newspapers, and by my wife.

None of these many lecturers was anything less than expert.

But now I have received the most severe, the most blistering, the most complete lecture of all.

It left me completely beaten, broken, bruised and brainwashed.

The lecture was given by my 2-year-old daughter. She has the face of an angel, but has the tongue of a viper. She has the most beautiful smile in the world, but the worst temper.

She was crying during the course of an argument with the twins. I told her to be quiet. Instead, she began to scream.

So I spanked her and told her get in her bed until she was quiet.

She marched off, screamed in her bed for about a minute, then came out, her face calm, the storm ended. She began to play with the twins as if nothing had happened.

But she hadn't forgotten the indignity of that spanking. A little later she caught me alone and she gave me the lecture.

Coming in baby talk, from a small and pretty girl who looks like a doll, the words were more effective than if they had been bellowed by a non-com.

"When you panked me I went to my room and I scweemed louder than ever," she announced, showing that the punishment actually had caused more, not less, noise.

"If you pank me again, I'll come in your bedroom and scweem loud here and you can't sleep good," she said, switching to a threat.

"When we go knock on doors, I'll tell people you panked me and they won't give you any candy."

This came from memories of Trick or Treat, and was warning of punishment for me if I got out of line again.

"I'll go see my sister Patsy and tell her you hit me." This is her older, married sister who lives in Oklahoma.

By this she was putting me on notice that she would get me in bad with the rest of the family, making my discipline look like outright violence.

In other words, like all women, she would put me in the bad with the rest of the family, making my discipline look like outright violence.

In other words, like all women, she would put me in the wrong by a slight distortion which she firmly would believe was true.

"When I gwow up, if yo pank me, I'll fight you."

This was plain intimidation.

"When I gwow up, I'm going to buy lots of things for mommy, and lots of things for Carl and lots of things for Evellen, but you won't get anything, not even gum.

"Daddies shouldn't pank little girls. I don't like to be panked."

Then she crawled in my lap and began to tease me by unbuttoning my shirt. Before she crawled down, to go looking for another argument with the twins, she kissed me.

So I'm forgiven for this time. But I'll be afraid to ever spank her again. I couldn't stand another such lecture.

Ever Run A Home?

There may be times when you think it is hard to get along with your wife, but you can take my word for it that it's much harder to get along without her. In fact, it's impossible.

The other morning my wife had started to prepare breakfast for the children when the grease in the skillet caught fire. She grabbed the skillet and started for the back door, but as she prepared to throw the skillet out, she tripped and much of the scalding grease spilled on her right hand and wrist.

Her cry of pain awakened me and I put some baking soda on the burn. Later I learned that this was not the best thing, and that I should have used petroleum jelly (such as Vaseline or other similar products) or shortening, but it seemed to relieve the pain.

The twins were up but not dressed, and the five year-old girl still was sleeping. My wife said she could drive the car, and in a frantic way we decided that would be better than to try to take the children to the hospital emergency ward. I should have called a cab for her, but I didn't think of that until she had driven away.

As soon as the hospital called to say my wife had arrived, I tried to be the calm, efficient head of the house.

"We will now get organized," I said.

"Wait until I get Miriam to see us organize!" shouted Carl, which he did.

"Everyone sit at the table and I'll get something for breakfast," I said.

"Mommy always has us dress first," said Evellen.

"Of course," I said. "I meant we'd sit at the table after we dressed. Now, everybody get dressed."

"Mommy has us wash our hands and face first," said Miriam.

"Naturally," I said. "Anyone knows that. And wash your teeth."

"Mommy has us wash our teeth after breakfast," said Carl.

After awhile, we finally got around to dressing. I told them that I knew they were old enough to do this without help, and they agreed. But within a second Evellen asked, "Where are my pants?"

"How do I know where your pants are?" I asked. "Don't you have a place for your pants?"

"Yes," she said, "but Mommy always finds them for me."

"I can't find them when I don't know where to look," I said. "Go find them yourself."

She looked at me with some disgust, but she found her pants. I then went through the same dialogue with Miriam, who couldn't find her shoes, and with Carl, who couldn't find his socks.

When they realized that it did no good to ask me where to find their things, they managed to find them, but you could tell that this wasn't normal procedure.

"Can I pick out my own dress?" asked Miriam.

I told her that she could, and she came out with a fancy dress and started to put it on, looking at me all the while.

"You aren't supposed to let me wear this dress," she finally said, "Mommy never lets me wear it to school. It's for special times."

"You can't wear that dress," I said.

Content, now that she had been properly denied the dress, she brought out another.

"This is the dress you are supposed to say I can wear," she said.

"I'm ready to eat," Carl announced.

"Very well," I said. "We will all sit at the table."

"I'm not dressed yet," said Evellen.

"OK, we'll start with Carl," I said, sorry that my organized schedule was working out so poorly. "What will we eat?"

"Since Mommy scorched the bacon, you'd better get us cereal," said Carl. "I'll get it for you." Which he did.

"Don't you know that Mommy always gives us milk and sugar on our cereal?" he asked.

"Naturally," I said. "I wonder where she keeps the sugar."

"I'll get it," he said, and he did.

I have run out of space, so I can't detail the rest of the way I got the children ready for school, but under their expert supervision, I did it.

When my wife came home, she asked if I had had any trouble.

"Not a bit," I said. "It just takes organization."

But if I were ordering a home, I'd rather have a one-armed wife in charge of it than a 10-thumbed husband.

We Have a New Driver

We have a new driver in our family.

He is a stranger who moved in with us about a year ago. About the time he came, our teenaged son disappeared. We have never heard from him since, but since he hasn't written for money and we haven't heard from the police, I assume he is getting along all right. At that same time, this newcomer moved in, pretending he was our boy.

He didn't fool me, of course, although I didn't let him know that I realized he was an impostor. Our boy was a skinny, wiry kid, but this stranger was a tall and muscular man, with a massive chest, bulging biceps and hairy legs that accounted for about three-fourths of his 6-foot height.

If his size hadn't given him away, his voice would have. Our little boy spoke in a high, loud voice. This fellow speaks in a low, loud voice. Every now and then he apparently remembers that the boy he is impersonating had a high voice and he will switch from his own bass to a falsetto, sometimes in the middle of a word. This is not a convincing performance, however, and I guess he realizes the futility at such an attempt to fool me. I notice he uses his natural deep voice more and more with fewer breaks into higher notes.

We have been going along with this impersonator so long now that I am used to calling him "Carl," the name of our son. I find it hard looking up to a 6-foot man while pretending he is my 16-year-old son, but I am getting accustomed to it. Still, every time he comes through the door or I see him on the street, I have a hard time convincing myself that this tall, handsome stranger is supposed to be my boy.

I will say this, he is the kind of adolescent young man I would have wanted my own little boy to grow up to be. If it hadn't happened so suddenly, almost overnight, I would even accept the fact that this giant is, indeed, my own offspring, the same one I used to toss in the air and catch while his mother protested I was apt to drop him and break his neck. It is disconcerting to realize that he could toss me in the air now, a thought I hope he never has.

But what I started to mention was that Carl took his driver's test this week.

He has never shown any feeling when he passed a subject in school, and he has hidden his emotions when he won a trophy or any other recognition. But this was the supreme achievement of his life. He couldn't disguise his pride. And although he usually doesn't speak to other members of the family except in grunts, he couldn't contain himself after this triumph and he had to announce it to one and all.

For him, of course, it is a mark of manhood. I couldn't bear to tell him that I didn't share his enthusiasm. I think in terms of increased insurance costs, the inevitable battle between him and his mother over use of our one car, the dangers of the road, fear of hot-rodding and one-arm driving and showing off to the girls, and drags, and chicken, arguing with other drivers – all the worries every father must have when his boy first gets behind the wheel and drives off alone, supremely confident and full of the joy of living, the feeling of a cowboy riding high in the saddle, the young eagle swooping from the cliff into the clouds.

Next chapter, I suppose, will be the jalopy or the wreck which he will try to repair or rebuild, and I will try not to think about that period until it arrives.

I have become so fond of this tall young stranger that I want him to live to a ripe old age and give us more Thompsons to hang in the family tree. (I don't think I said that the way I intended.) I hope he realizes he now has been authorized to handle something that can be far more dangerous than a loaded .45.

In fact, if he'll just stay out of accidents and not collect traffic tickets, I think I'll accept this tall young man as a permanent part of the family and quit expecting my little boy to come back.

On Owning a Junkyard

I don't know if we'll ever buy that second car, but at the moment it looks as if I might wind up owning a junkyard, or at least a yard for Volkswagen spare parts.

My boy Carl located this Volks with detached engine and then it turned out that one of our friends is a Bug bug. (Volks owners, I find, fondly refer to their cars as 'Bugs.' At least, I suppose it is used fondly.)

The Birth of an Auto

Not since the first Model T Ford rolled off the assembly line has there been such a big step taken in the automobile industry as the birth of a second-hand Volkswagen one day this week.

The parents were an old Volkswagen with an engine that wouldn't work and an unworkable engine from another old Volkswagen. With Joe Musial as the chief midwife and a corps of volunteer helpers led by my 16-year-old son, the labor pains extended over a week's period during which the best parts of the two engines were married, as the engineers say. The resulting offspring was a 1960 model Volks with a rebuilt transplanted engine. With a genius exceeded only by that of the original Ford, Musial had created one workable engine out of the two unworkable ones.

If I didn't believe in the mechanical genius of Mr. Musial, I would call it a miracle. And Carl thinks it was just that.

When Mr. Musial told him to get in and turn the ignition and the motor began to cough loudly, Carl said, "Hear it purr!"

His face beamed so brightly he looked like a full moon.

Surgery on an Old Car

After the initial road test revealed that the only thing noticeably wrong with the rebuilt Volkswagen was an empty gas tank, we bought it. The lady at the tag agency where we had to transfer the title and pay the sales tax looked dubious at the price, since it seemed incredibly low. She didn't know that it looked like a pile of junk on wheels in the original version and she would have been incredulous about the high price. In its rebuilt, running, condition it did look like a real bargain, but no one could tell about the many hours and extra engine that had been donated.

And so the car was ours, and with trumpets and drums sounding in his mind, Carl once again prepared to drive around the block.

The car made it without losing any parts, but Carl drove several more blocks, back to the factory, the Back Yard Volkswagen Rejuvenation, Rehabilitation, Restoration, Rebuilding and Renovating Company Unlimited, where he reported to the president personally. He said that a red light was warning that the oil was low and that he had noticed a puddle of oil under the car that morning.

Joe Musial, the president, shop foreman, chief mechanic and grease monkey at the factory, figured Carl had driven the car after it ran out of oil, and he first feared stripped gears, but the stick showed oil, even if it was circulating properly.

This called for a consultation. After consulting with himself for some time, Joe thought of half a dozen possible explanations. He decided finally that an exploratory operation was necessary.

This is major surgery but not fatal in itself. Still, the average car owner is afraid to have one performed on his car. The price of the surgeon is astronomically high, for one thing; it is not certain that the cause of the ailment will be discovered; if it is, there is no guarantee that it can be cured.

Also – let us face it – many otherwise competent car doctors have trouble putting a patient's parts back together once they've taken them out of the body. In such cases, the operation itself can be the cause of death.

However, since we had unlimited faith in Dr. Musial, and also because he was the only free car doctor we knew, we gave him the go-ahead signal.

Musial finally got the engine back together and crammed inside the body and pronounced the operation a success.

Between rains and the fact that my boy was working at the market after school, it was that weekend before Carl went by to pick up the chariot.

Again the bugles bugled, the trumpets trumped and the drums drummed in his mind as he took the keys and sat in the throne behind the wheel. The car wouldn't start.

Score to date: Two road tests, two breakdowns. Present status: Car back in factory for checkup; Carl's faith undiminished, my faith turning to skepticism, Musial looking as if he wishes he was back in the Army, calvary preferred.

Casualties

Sort of a (Gulp) Lark

Sometimes at night I would wake up with my tongue hurting, which seems like a funny thing to wake up with. I have had my tongue for many years, and it has never done that before.

At first I thought that maybe I was biting my tongue in my sleep, but this didn't make much sense, and so one day when I was seeing my doctor I mentioned my tongue to him.

He took a look at it and right then I discovered that there is nothing more awkward to show to somebody than your tongue. They can put a stethoscope to your chest and listen to your heart or they can take X-rays and study your lungs. They can look in your eyes and if you break an arm, there the arm is for them to see. If you have an earache, the ear is sticking out ready for them to peer into.

But a tongue simply doesn't lend itself to examination. And while I gagged, wishing I hadn't brought up the subject, my doctor stuck his nose in my mouth and said a few hems and haws.

Obviously he didn't see anything but a tongue there, but he didn't want to admit that, so he referred me to a specialist. Several weeks passed before I kept my promise to call this specialist, but I finally made an appointment.

Once in this doctor's office I was subjected to the same unhygienic examination, made worse in this case because this doctor insisted on running a finger along the edge of my tongue. I tried to protest, but all I could say was, "Uggg, ugggg!" which he interpreted to mean, "Good show. Keep it up." So he kept on running his finger around on my tongue and when he finished he said he had found a lesion on it and that although he didn't think it amounted to much that I ought to have it scraped off right away.

Since I already owed him for the office call, I figured I should follow his advice, so I said okay, start scraping.

But he said I would have to go to a hospital and that if this wasn't any more than he thought it was, it would be a minor thing – sort of a lark for both of us.

I said fine, but would I be able to go to work the next day, and he said, well, now, if it was what he thought I should be able to leave the hospital that day, but on the other hand if it was worse, maybe he would have to scrape a little deeper and that might delay me slightly.

I said, okay, when do I go to the hospital, and his nurse reserved a room for me just as if I were going to a Miami Beach hotel and she reserved the operating room, and I got out of there thinking what a smart man I was to have this little thing taken care of so soon.

I had one day left at the office, and at first I said nothing to anyone about my impending little tongue-scraping.

Finally, I did mention it to Lil, the secretary, who is the real boss of our news room, and she was greatly impressed.

She was so impressed, she impressed me.

"You mean they are going to operate on your tongue?" she gasped.

I said yes. After all, even if they just scraped it, that would be an operation – and it was going to happen in the operating room.

"Is it – is it anything serious?" she asked.

"Well, let's just say I hope for the best," I answered.

I felt very brave when I walked away from her. The next person I stopped, I said, "Well, I have to get my tongue operated on tomorrow. Nothing serious – unless it's malignant, of course. There's a chance it won't be. I'm not worried." And I walked off whistling, wondering if he didn't consider me the bravest man of the generation.

By my third person, I could hardly get through a sentence.

"I probably won't be seeing you for awhile. They're operating on my tongue." My eyes filled with genuine tears. "It may not be cancer," I stammered. "But you never know."

I was practically a basket case by the end of the day. Now I not only worry about survival, but I have visions of going around with my tongue in a cast or a splint. How can you do that? How can they even wrap your tongue in a big bandage?

It started out as a minor scraping job, but now –

Well, if I get out of this at least it has taught me a lesson. I'll never stick my tongue out to a doctor again.

She Took My Clothes

The first thing they do to you when you go to a hospital is make you feel completely helpless. Before they are finished with you, you are so debased that you accept any degradation without whimpering.

I had walked into the hospital on my own two feet, a reasonably robust appearing man for a person of my age and condition, but when I was assigned a room, they told me to get in a wheelchair.

I protested that I was perfectly capable of walking to my room, but nobody heard me.

My indignation was somewhat diluted when Julio Bueno appeared to push my wheel chair. He is the aide who had wheeled me in and wheeled me out when I was in this same hospital slightly more than a year ago.

Julio began to push me up the long corridors.

"You've gained some weight since last year," he observed. I felt like telling him that maybe he had put on some weight himself and shouldn't blame me just because I seemed like a heavy load, but I already was accepting the hospital philosophy that a patient knows nothing.

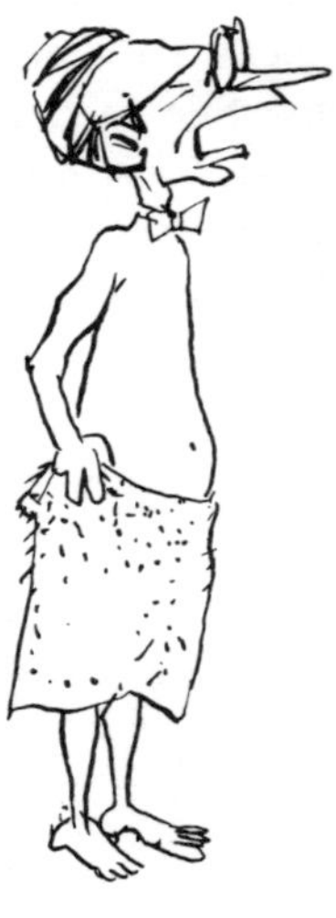

When we reached my section, there at the nurses' station was Mrs. Nancy Weir, who is in charge of the floor, and several other of my old friends from last year. For a fleeting second, as I was wheeled by and we waved at each other, I felt like a home town boy returning to a hero's welcome. Then I remembered why I was there and quit feeling like a big shot and began to feel sorry for myself.

Sure, I told myself, naturally they were glad to see me. They now have a new specimen to work on.

When we got to my room I got out of the wheel chair, walked over to an easy chair and was looking out of the window when Mrs. Thomas came in.

I didn't know it at the moment, but this was the beginning of the battle of the drawers, between Mrs. Thomas and me.

Seeing that my ride in the wheelchair hadn't completely broken my morale, it was her duty to strip me of my dignity – or, strictly speaking, of my clothes.

She told me to take off my clothes and get in bed. I told her I felt fine and since my operation wasn't until that afternoon I thought I'd sit up for a few hours and read a magazine. She repeated that patients undressed and got in bed.

"You will wear this," she said, holding out one of those ridiculous short sheets which hospitals call gowns.

I undressed, down to my drawers, and I put on that silly gown which covers you in front from your neck to a point half-way between your hips and your knees. She tied the strings at the back, around my neck and chest. This leaves practically all of your backside exposed. If you wore one of those things to the beach and stayed on your stomach, you would get sunburned in places that would keep you from sitting down.

Anyway, I submitted to this indignity, then I jumped in bed, bottom down, and pulled the sheet up to my chin.

"Now," she said firmly, "let me have your drawers."

Frankly, I was embarrassed to think that a nurse would talk to a married man and father of five children, not to mention two grandchildren, like that and I made a mental note to report this to somebody, but I shucked off my shorts. I pretended to close my eyes, but I peeked as she put my clothes away, and saw that she put my drawers in a dresser drawer.

Then Mrs. Thomas began to fill out forms. It takes more forms to get in a hospital than it does to get a government job.

She wrote down, and recited to me, a list of all my clothes; noted that I had one pair of glasses; a $1 bill (which I had brought along just in case I felt like any reckless spending during my day's vacation); one pencil which I had stolen from The Herald.

"Let me have your wrist watch," she said.

"I want to know what time it is," I said.

"Would it be all right if I gave the watch to your wife instead of locking it in the office?" she asked. "You can get it after surgery."

The way she was itemizing my estate and distributing my assets, I began to wonder if I was expected to return from surgery.

"Now," she said, "sign this."

Vaguely feeling it was my will, I signed. By that time I had signed so many things that one more couldn't hurt anything.

I Was a Guinea Pig

Word spread rapidly in the hospital that they had a new guinea pig imprisoned in one of the rooms.

As soon as Mrs. Thomas had undressed me, the registered nurses, the practical nurses, the aids, the interns, the Candy Stripe girls, the student nurses, the plumber, the janitors, and the staff doctor began to visit me. I am not sure, but I think a few tourists dropped by. They may have been selling tickets.

Apparently whenever anybody connected with the hospital had nothing else to do they would stick a thermometer in my mouth.

One lady did this and left for other duties. Half an hour later, I buzzed the buzzer and another nurse removed the thermometer. Half an hour after that the first lady returned. She looked like somebody who had lost something.

"Didn't I leave a thermometer in your mouth?" she asked.

"Yeah," I said. "It dissolved and I swallowed it."

That didn't upset her. She just stuck another thermometer in my mouth. This time, to make sure it didn't disappear, she stayed and removed it.

During one of the few periods when nobody was in the room, I sneaked out of bed and got my shorts out of the dresser drawer. I felt much better when I had them back on.

Somewhere along the line they had started to give me sleeping pills, and then Mrs. Thomas came back with a long needle.

She told me to assume the position, which means the needle is not going in your arm, and I rolled over.

"You put your drawers back on!" she said.

"I'm afraid you don't understand what they are going to operate on," I said. "It is my tongue. I don't need my pants off for that."

"Take them off!" she said. "Nobody wears underwear in the operating room!"

So once again I took off my drawers and once again she put them in the dresser drawer. Then she jabbed a foot-long needle in me.

They thought they were putting me to sleep, but I was too smart for them. I wanted to know what they were doing to me.

Once, before they had doped me too much, I decided to take care of a personal problem while I was able. Again I sneaked out of bed. I peeked out of the door and there wasn't a person in sight.

So I walked across the hall to the men's room, holding the back of my nightgown together with one hand.

When I returned to my room, a nurse was in the corridor. I walked with the dignity of a man on Fifth Ave. on Easter in his top hat, ascot and striped pants.

It is not easy to act that dignified when all you have on is a short sheet that covers only your front and reaches only halfway to your knees and you have to hold the back sides together.

The way that nurse stared at me, I must have been the most dignified looking person she ever saw.

Back in bed, I was subjected to more experiments.

One nurse took my blood. I didn't look, having a prejudice against seeing my own blood, but apparently the blood bank was running short. She took approximately a gallon.

They sent a man in for another kind of sample which had nothing to do with my tongue, and another nurse wrapped my arm in a used inner tube and took my blood pressure.

Somebody else used a stethescope to see if I was alive, and a staff doctor came in to fill out a stray form he found in his files, asking such things as how many operations I had had in my life, whether I drink hard liquor, if I preferred blondes to brunettes, and if my life insurance was paid up.

I was getting so groggy by then I was not sure about the questions, but that left us even. He couldn't be very sure about my answers, either.

Finally, I did go to sleep, but every now and then they would wake me up to give me another sleeping pill and I think that Mrs. Thomas slipped in to give me another needle, but I am not sure about that. Maybe she just came in to check on my drawers.

There sometimes are stays of executions and the sentences even are changed, but I had already signed my rights away and there was no hope for a reprieve or a postponement of my operation.

At 12:30 p.m. they woke me up and prepared me for that Last Mile.

Electric Chair Episode

Since I have never been executed by electrocution, I am not sure about the ritual, but it seems to me that they shave a man's head and sponge him liberally with salt water at various places where the electricity is to make contact.

Anyway, when it was my time for surgery in the hospital they gave me approximately the same treatment; so close, in fact, I wondered if I was in the wrong institution.

They sponged me all around the mouth with something that was supposed to freeze my face and then, instead of shaving my hair, they covered it with a hat.

The only thing I can compare this to is a shower cap, like women wear. It has an electric – I mean elastic – band and they pull it down until it covers the top part of your head so that you look like something the cat dragged in.

Actually if a cat of mine dragged in anything looking like that, I would fire the cat.

The more I think about it, this hat looked more like something a scrub woman might wear.

Then two men brought in a stretcher on wheels. They wanted to put me on that, but I was able to roll over and get on it without help. I was pretty proud of myself for that small display of independence.

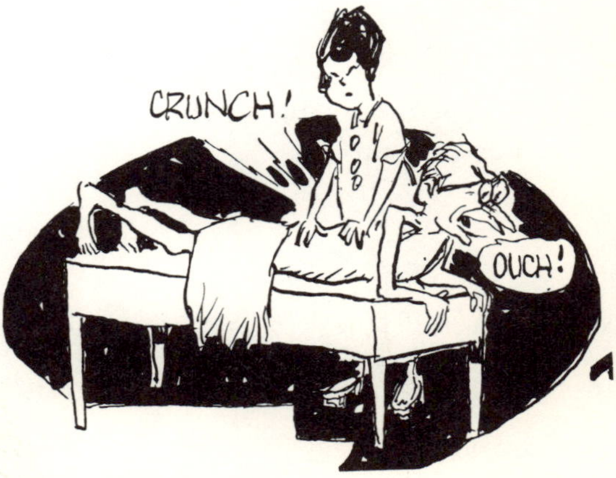

Then they began to wheel me up the corridors. They must need a map to find that operating room, the way they kept pushing me for miles and miles.

I felt just like a patient in a Ben Casey program or the star of a movie. I imagined that nurses and doctors and possibly hundreds of admirers were jammed along the corridor watching me as I was wheeled along, saying such things as, "Doesn't he look heroic?" or "What a handsome man!"

But once I managed to hold my head up enough to look around, and nobody was watching. Besides, I got a look at myself in a mirror, and the way I looked in that scrub lady's hat took all the glamour out of the whole deal.

It was then that I realized they meant business. All of a sudden, I decided that the trip wasn't necessary, and that I wanted a few more days to think it over.

But that is like deciding you don't want to take a roller coaster rider when your car reaches the top of the first dip. It's too late to change your mind.

We finally reached the operating room. I was determined to notice everything that went on and I raised my head. There were swinging doors.

I can't remember now if the sign on the doors said "Surgery" or "Operating Room." It doesn't really matter, but it bothers me that I turned out to be such a poor reporter.

Once we got inside, there was no doubt about where we were. We were in the execution chamber – I mean, the operating room.

It wasn't a very big room and practically no furniture except for a table in the center of the room, directly under a huge, round light.

They wheeled me up to this table and once again I rolled on to it by myself and once again I was very proud of this achievement. Maybe I was trying to prove that I was so healthy that I didn't really need an operation.

There was no question about the fact I was the star of the show; the main guest at the party. The doctor was waiting for me, his nurse was waiting for me, other nurses were waiting for me. For some reason this didn't make me very happy.

Directly above me was this big light. I counted carefully, to make sure I would remember. There was one big bulb in the center. It was surrounded by three or four other big bulbs.

It is embarrassing that I can't remember how many there were, because I counted them at least a dozen times.

Then the doctor got on one side of me and a nurse on the other. All I could see was their faces and since they wore masks, all I really could see was their eyes. I would say their eyes were just so-so, hardly worth all the trouble I had gone to just to see them.

They put some padding in my mouth. Apparently they tore a Herald into shreds and stuffed it in. It was a Sunday edition.

Then they put in some cotton. You have seen these miniature bales of cotton they sell as souvenirs. They rammed one of those in my mouth, next to the newspaper.

Then the doctor asked if I was ready. Naturally I couldn't say anything, but for some reason he got the idea I had said yes. I am not sure, but I thought I heard the sound of knives being sharpened on a grindstone. Everything was ready except me.

Whose Tongue Was It?

I had intended to give a detailed account of my operation, but although I kept my eyes open, about all I can truthfully report is that the big clock in the operating room said 1 o'clock when I was wheeled in and that it said 1:30 when I was wheeled out.

Since I missed about everything that went on during that half hour, I doubt if I will win a Pulitzer Prize for my observations.

In givine me local anesthetic, they did stick a couple of needles in my tongue, and although this makes me cringe as I think of it now, at the time it didn't feel any worse than the needle they stick in your arm or elsewhere for a vaccination or a hypo.

To tell the truth, the operation was rather boring. Despite my keenest efforts to see what was happening, all I ever saw was those two heads, with eyes showing above the masks. This is not my idea of a thrilling show.

Once I did hear, or thought I heard, the doctor say, "Scalpel!" and that reassured me. I have seen enough of Ben Casey and Doctor Kildare to know that no operation is legitimate if the doctor doesn't ask his nurse for a scalpel. Sometimes I must learn what a scalpel is. I think it is something you use to take poached eggs out of the water, although that might be a spatula. They probably are the same thing.

But while I was lying there, my mouth open like a fish out of water, completely helpless, I did think of one thing.

One of those faces above me belonged to a doctor I had never seen until two days before. Now my life depended on him. I had placed that life in his hands. I had agreed that he should do what he felt was necessary for me. My own doctor had referred me to him. I had confidence in my own doctor; I now had faith in this doctor.

This doctor now was my own doctor, and even as he was cutting out a portion of my tongue, it never occurred to me to doubt him.

I wonder if medical students realize the awesome responsibility for which they are training; if interns fully appreciate the importance of deserving the faith that patients must have in their doctors; if all practicing physicians are fully aware of the fact that they must win and merit the complete confidence of those they treat.

A doctor who does not have and merit such confidence is not a good doctor; a patient who does not have such faith in his doctor is a pathetic person.

I had such faith in this stranger who now was the most important person to me in the world. It never occurred to me to doubt his ability, his interest, his integrity.

He finished with what he was doing and he rushed me off to the laboratory with what he had cut out of me, and his nurse remained, fingers in my mouth. As I say, I was trying to be a good reporter, but I could not tell how many fingers she was using.

My mouth was pretty full at the time, even with part of the tongue gone, and I was numb so one finger or five would have felt about the same.

She kept up a constant stream of small talk, and I had enough sense to realize she was trying to keep my mind off the only important thing in the ball game: the test being made in the laboratory.

Once when she asked me some question, knowing full well I couldn't answer, I decided to see if I could bite her fingers. I didn't know if I could work my jaws, but I tried and I was very pleased when she let out a yelp.

"That hurt!" she said, and I did my best to laugh, which is hard under such circumstances. I still think it was funny.

The doctor came back, literally running. "Good news!" he shouted. "It's negative!"

"Did you hear that?" the nurse asked. "It's negative! You're all right!"

I didn't know what she expected me to do about it. I was strapped in, doped up, and my mouth was stuffed like a Christmas stocking. I tried to clap my hands but they were under a sheet and it didn't work.

"This is my lucky day!" said the doctor.

I felt like asking him whose tongue he thought he had operated on; it was my lucky day, not his, but I was in no position to argue. Besides I was too happy – and sleepy.

A DAY IN MY
LIFE AT THE OFFICE
AFTER A FEW HOURS AT MY DESK THIS IS RELAXING
STUMPED ON·AN IDEA FOR A COLUMN MR.T. CAN ALWAYS GAZE ACROSS BISCAYNE BAY AT THE MIAMI BEACH SKYLINE..

Goodbye, Old Building

Please excuse me if I keep talking about moving to the new Herald building, but this is about all I can think of right now.

This is the last day in the city room in the old building, and if the teletype operator has trouble setting this copy, it's because of the tearstains on the paper. (It also might be because I have a worn-out ribbon in my typewriter and because I misspell so many words it is hard to tell what I am writing.)

The notice on the board says that after today we all report for work at the new building. That seems very definite.

And though I am the last one to start packing, I have started preparations.

They have provided us with cardboard boxes of all sizes and shapes. I have taken one of these and have it beside me now. It is yawning at the top, its big mouth begging to swallow all the things I have hidden in my old desk. I am reasonably sure I could never sort out this vast and ancient accumulation, but I have a solution.

When I have finished work today, I will remove the drawers from my desk, empty the contents into the big box. Then I will crawl into the box and pull down the top.

At last, I will be ready. The rest will be up to the movers.

I Want to Stay 'Home'

I guess The Herald is serious about moving out of the building on S. Miami Ave. into that super-colossal structure on Biscayne Bay between the causeways.

I know that they have been talking about that new building for many years; I know that it has been materializing before our eyes and that it was announced weeks ago that the time was near for the moving.

Somehow I really never believed that it would happen, that we actually would get out of this building, which is the only Herald home I have known since I came to work here in December, 1945, a building on the only site The Herald has ever had, since it was started in another December, in 1910.

But we are moving. As this is written, we are almost moved. Only a few departments remain at the old stand.

When I came to work this morning, the first floor business office, usually crowded with ad takers at the counters, ad salesmen at their desks, secretaries and executives, was almost as barren as a ghost town.

Up on the fourth floor, things looked even more deserted and desolate, because big holes had been torn in walls to facilitate the removal of heavy machinery and entire offices are stripped to the bare walls, as if the janitor had just cleaned up following a raid by vandals.

Between the first and fourth floors, things are in a stage of limbo. On top of or beside each desk there are big boxes for packing, and the typewriters, filing cabinets and other furniture to be moved are prominently tagged, showing where they will be delivered.

I know – now at last I believe – that we are leaving our old home.

And the funny thing is that I am sad about it.

Almost as soon as this building was completed, in 1941, employees began to complain that it wasn't big enough. These complaints continued even after a four-story addition was completed in 1946.

Certainly, in the 17-plus years I have been here there has been a constant chorus from the news room about the need for more space, and I was one of the loudest members of the chorus. We complained that the building simply wasn't adequate to take care of the phenomenal growth of the area and the equally phenomenal growth of the newspaper.

Now I am ashamed of my complaints.

I look around the city room and I think of the stories I have written here, the people I have interviewed here, the people I have worked with, and I find that I cannot disassociate them from the crowded, noisy room.

Over on the copy desk is Al Hallman, a veteran of many newspapers, sitting on the rim with those other war horses of the old school, Clarence Good, Jess Irwin and Roy Kelly. Bob Frederick is amusement editor, Art Peavy is a star reporter and Ernie Hill is covering Latin America.

In the corner which now is the managing editor's office is Arthur Griffith, that fine gentleman, writing editorials, while Abe Hurwitz is helping on the city desk. On Sundays, Henning Heldt, our political writer, is the city editor, and a few years later the city editor is Merlin Test.

In the woman's department, Eileen Bryne, later to become Mrs. Lee Hills, is the fashion editor, working under Mary Schuck Smith, the woman's editor. Ferman Wilson is the real estate editor.

Downstairs, Lee Gash is head of the press room. Earl Thomas and Martin Tracy are among the ad salesmen. The Earl brothers – Fay and Guy – are in stereotype, and Harry Clifton in engraving, Walter Anderson in composing.

They are all dead now – these and many more – and this is the building that contains their memories.

This, of course, is the building where the papers were printed telling about Pearl Harbor and the defeats that followed, the long way to V-E and V-J days, the stories that helped break the back of organized crime and won a Pulitzer prize, the building that saw the birth of the Clipper edition and civic campaigns that helped a growing city be great as well as big.

Into this news room have come the woman with the goat, the man with the boa constrictor, the man with a donkey, the man with the chimps, Babe Ruth and Jack Dempsey, Fred MacMurray and Jerry Lewis, the famous, the infamous, the great and the ingrates, statesmen and simpletons.

Please, Mr. Knight, can't we rent out that new building and stay here a few more generations? Belatedly, I say it. This is a fine building, a wonderful building, a building of history – and memories.

Pressdom's Palace

It was a bright and breezy Sunday afternoon when I went to work for the first time in the Taj Mahal of Biscayne Bay, the mammoth, majestic, monstrous, marvelous, magnificent new Herald building.

Although my wife drove me and could have let me off at the front doors, I had her go into the employees' parking lot, to give it sort of a test run. I was sorry the gate was open, so I didn't get to use my card. Parking our cars is like belonging to a club. We insert this membership card in a machine and that opens the gate.

Unfortunately, I didn't get to see this mechanical marvel work, so I'm not entirely convinced yet that it will.

We drove around the parking lot and figured that if I ever had to park at the end farthest from the building, I would have to catch a bus to reach the door the same day. Maybe that's an exaggeration, but I'd hate to hike the entire distance of the lot carrying a full pack.

Then we tried to get out, but we missed the exit gate the first time around and had to make a complete tour again. By now this was running up quite a gas bill.

We found the exit gate on the second try, and I got out of the car and walked up to our new home. As I approached, I had a horrible feeling that I might be met by a butler who would send me home for white tie and tails.

I got in the lobby without being challenged, but once inside the lobby I was lost.

There was a man in fancy uniform, even with a pith helmet, sitting near the escalator.

"Could I help you?" he asked.

"Why yes," I said, "I'm trying to find the city room."

"More people have been asking for that," he said, "but we don't seem to have one. Could you use any of these other departments?"

He pointed to the large directory on the wall.

"Here I am," I said, pointing to "News Department, Fifth Floor."

"I thought you said city room," he said, looking put out.

"Now how do I get there?" I asked.

I felt a little ridiculous, a Herald employee for going on 18 years, asking an obviously new employee how to get to my desk, but the young man politely gave me directions. I had my choice of escalator or elevator, and naturally chose an escalator. I never saw one of those things until I was a grown man and I still think they're exciting.

At each floor I would transfer to another escalator and I kept riding and riding until it seemed to me I surely had run out of building, but the roof was still above me.

Finally I was convinced I had gone past my floor, but before starting down, I wandered up the hall, through a door – and there was the city room, big as a football field. It is so modern it even has clothes closets to hang our coats in.

I found my desk, at one end of the field, and my good old typewriter provided a comforting link with the past. It was good to find that it worked the same – no better, but no worse.

My desk is a fashionable green in two tones, with chairs to match. Come to think of it, all the other desks in sight are the same colors, except for the women's department, to my back, which are a prominent yellow.

I gingerly tried out my new chair and prepared to write a piece for the paper. Automatically, I reached for a cigaret. Before I lighted it, I was struck by a horrible realization.

There was no ash tray on my beautiful new desk! I am no fool, at least not a big enough fool to put a lighted cigaret on my beautiful new desk the very first day I am working in our beautiful new building. I went running for the managing editor and he quickly supplied me with the necessary tray. He told me that I had done the right thing, indeed, not to light a cigaret without such a tray handy.

On weekdays, I will be sitting near Jack Kofoed, Jack Bell, John McDermott, Adon Taft and assorted other human beings, but this Sunday afternoon I was alone at my end of the floor, and I felt lonesome.

It occurred to me that I hadn't given my wife my new phone number. Then it occurred to me that I didn't know my new phone number. I decided to ask the city editor.

I picked up my pretty new beige telephone. "Long distance," I told the operator. "Get me the city desk."

Lost: Composing Room

Adventures on the First Day in a New Building, Second Chapter.

On that First Day, while all of us were feeling self conscious and like intruders, there was an explosion in the northwest section of town. The city editor sent Stu Auerbach, a reporter, out to meet a photographer and cover the story.

Auerbach, who is built roughly like a fire horse, Percheron type, leaped from his brand new chair, flew from his brand new desk, grabbed his coat from the brand new clothes closet, and dashed madly toward the end of the mile-long room.

Some time later, he came charging back headed for the other end of the field, still holding his coat in one hand, the tail standing out stiffly in his wake.

He made it half way to the other end when he put down his heels and ground to a smoking halt.

"How do you get out of this room?" he asked in a quiet voice.

Soon after a friend from composing room wandered by my desk.

"Sure a big building, isn't it?" he asked.

"Sure is," I said.

"It's so big they lost the composing room," he said.

"Ha, ha," I laughed, to show I could appreciate a good joke.

He went on and many minutes later wandered my way again.

"No fooling," he said. "I can't find the composing room. Where is it?"

Naturally I didn't know, but it turned out that he was on the wrong floor.

One of the things I forgot in moving was a phone book.

When I decided to make my first call out of the building I discovered that everybody else on the floor had forgotten the same thing.

It was a job trying to reach our operators, for they were deluged with calls and were trying out the intricate new switchboard which eventually will make our hoseu one of the most efficient phone operations around.

Finally, I dialed to get an outside line, then dialed The Herald's new number, 350-2111, thus reaching one of our harried operators. By then I had forgotten what I wanted, but I found that my own number is 350-2664. Call that and you get me direct, if I am here. I can hardly wait to get home so I can call myself.

Nothing surprises you about a new building. I hardly noticed it when a couple of men went by carrying a tree, or at least a big plant, headed for Mr. Knight's office. Maybe he's going to raise a forest in there.

But the copy boy was surprised when he pulled on the knob to open the door that would let him out of the city room – and the knob came off in his hand.

And the teletype operators were surprised when the new conveyor belt which brings copy from the copy desk began to spew papers all over the room.

These things, of course, are the minor troubles that must be expected in a new building. Long before our open house, Sunday, April 7, when you should come visit us, we should be so used to our new surroundings we can act nonchalant about them.

I hope by then I will have found a pencil sharpener, that I no longer will have to ask the girl on the information desk how to find the men's room, and that I have summoned enough courage to talk naturally on the phone instead of whispering.

Historical Item: The first classified ad to be placed in the new building was by Mrs. Hilda Herns, who wanted to rent a duplex. Knowing that Herald classified ads will lose none of their pulling power because of our change of address, I am confident she will find a renter.

More Unfinished Business: I have to find where the cigaret machine is located. There must be one somewhere in this big building.

The Hat of Journalism

It is an old university custom, at Oxford, Harvard and places like that, to dedicate a chair to somebody. This honor is reserved for great scholars or for rich men who give a lot of money to the school.

Well, modest as I am, I am forced to announce that I have achieved recognition much more flattering than to have a chair named in my honor.

It is with appropriate blushes and bows that I reveal that Miami Carol City High School has a Thompson Hat of Journalism. As far as I know, this is the first hat to be recognized by any educational institution anywhere in the world. An outstanding journalism student will wear the hat his senior year.

Debby Oney, of the journalism department of Carol City Hi, asked for the hat. I told her that I hadn't worn a hat since moving to Miami almost 20 years ago. She said it was awful if the first Hat of Journalism went unfilled, or even undonated, so I scrounged around until at last, in the bottom of a trunk holding old daguerrotypes and stereoscope pictures, I found one of my beaten-up old hats.

Into the sweat-marked hat band I slipped an old police card, void and with my picture ripped off, but still a press card, for no journalism student could wear a journalism hat without a press card. If you have ever seen a reporter in the movies or on TV, you know that.

In real life, the press card is of value when you are trying to crash a party, run a tab, or cash a check, and there are rare times when it helps you get by a volunteer cop at a parade. Its only other value is to get you slugged if you are covering a riot or other kind of mob trouble. Both sides always hate each other in such arguments, but they join in hating newspapermen, probably because they are afraid of the dispassionate, objective reports which rip the masks of respectability from men gone mad.

In such a case, the Thompson Hat might get a journalism student a lump on his head, but I trust that the Carol City journalists will stay away from controversial gatherings.

As a student journalist, I reveled in controversy, but my club was not heavy.

In high school, as editor of The Stillwater Stopper, I campaigned vigorously for more pep at the football games and for cleaner lockers in the halls. In college, as editor of The Daily O'Collegian, I campaigned vigorously against students making cow paths on the campus by not staying on the sidewalks, as well as fighting against the 7:30 p.m. curfew for dates on weekdays. I thought the curfew should be extended to 8 p.m.

I was crusading, a fighting editor, and I hope that the young man who wears the Thompson Hat of Journalism carries on in this fine tradition of harmless word warfare.

Now, of course, I did not wear this same hat when I was in high school or college. I wore it when I was an old war horse, a boisterous, cynical, hardened but soft-hearted brilliant, reporter of the Richard Harding Davis, Gene Fowler, Ben Hecht, Bob Casey, Walter Howey, Jack Lait school, a man who solved murders for breakfast and exposed graft in high places for lunch, reserving rescues of beautiful women from dope rings and fates worse than death for after dinner hours.

So, whoever gets to wear the Thompson Hat for the next school year, wear it with reverence and handle it with care.

It was in this hat that I took the wrong subway when I was rushing to interview the bombing victims at the hospital in the Bronx and ended up with a story about a new chimp at the zoo, a feat which won me no compliments from the city editor; it was in this hat that I couldn't find a phone to call in the big fire in Philadelphia until The Inquirer already had splashed it over two editions; it was in this hat that I got thrown out of a press conference for General Wainwright because my special pass fell out on the elevator; it was in this hat I let a pretty girl reporter grab the only available phone and scoop me on the fire at Palisades Park.

Yes, the hat has a history. It is a hat that has been lived in. Treat it with the disrespect it deserves.

Reporter's Life Tough

A reporter often is held personally responsible for the actions or statements of people he writes about, although he doesn't make the news but merely reports it.

My main job these days is to fill this space, but occasionally I write a news story, to justify my place on the payroll and to reassure myself that I still am a reporter.

There is a great deal of difference between a columnist and a reporter.

To write a column, about all you have to do is hit the capital "I" on the typewriter and say what you think about anything. It doesn't matter too much if you know what you're talking about, if you can make yourself sound important.

A reporter has a much harder job. He isn't permitted the luxury of a personal pronoun or the laziness of idle observation.

He must keep his opinions out of the story. If he isn't thorough, accurate, fair and objective, he isn't a good reporter. His job is to be an observer, not a commentator. He has to keep his eyes open and his mouth shut.

A good reporter is the most valuable asset of any paper. Sure, you need advertising salesmen and circulation men, too, but if the reporter isn't giving the readers an honest and a complete picture of what's going on, the advertising and circulation men don't have much to sell.

The reporter's job not only is important, but it is difficult.

At this moment I will skip the difficulties of getting a story, writing it, and then getting it past the city editors and copy desk, but will speak only of the troubles he has after the story is printed.

During a political campaign, every candidate is apt to put ads in the paper. Supporters of Candidate Jones don't blame the advertising department because they don't like what Candidate Smith says in his ad. They realize that it came from Smith, not from the ad salesman.

But if a reporter covers a rally at which Candidate Smith says the same things he put in his ad, and the reporter correctly quotes him, then supporters of Candidate Jones claim that the reporter has sold out to Smith. The fact that the same reporter correctly quoted Jones at another rally is forgotten – except for by Smith, who is sure that the reporter sold out to Jones.

The other day I wrote a story about four Edison High School students who want to go to Europe. They got a charter in Tallahassee, got the governor to sign it, and now plan to sell "stock" to finance their trip.

The next morning I got a call from an indignant lady who accused me of everything from dishonesty to immorality, because she said I had advocated that four kids beg money.

Now I had neither condemned nor indorsed this project. I simply reported what these students were doing and trying to do.

When I tried to point this out, the lady said that if I hadn't been trying to help them, I'd never have written the story. But if the story hadn't been printed, she never would have known about the project and couldn't have been indignant about it. Her very indignation proved that the story had been of interest. But she thought it was wrong for me to write it.

Another time I wrote a story about a lady who gave her deceased parakeet a fine funeral and an elegant grave, including headstone.

I got some nasty letters about that, from people who thought it was misplaced sentiment for a lady to make over a dead bird in such a way. All I had done was write about the lady and her parakeet, without personal comment. It wasn't my bird or my funeral. The very fact that these readers got so worked up indicated that the story was of interest – but they thought I was awful for writing it.

I have interviewed murderers and given their versions of the crime, but that doesn't mean I believed their stories. I have written of worthy programs, but that doesn't mean that I was responsible for them. I have interviewed saintly men who have performed great works, but that doesn't mean I deserve credit for their deeds. Unlike a column, a news story is not partisan, personal nor prejudiced.

It's easy to be a columnist, because you only have to express and support your own opinions. It's rougher on a reporter, who must accurately report the opinions and actions of others – and then be held responsible for them.

The Job of Waiting

As far as I know no school of journalism offers a class in "Waiting," and that is overlooking one of the most important phases of the profession.

Most young people who want to be newspapermen think of it in terms of exciting stories, stirring adventures, big headlines, tremendous scoops, rubbing elbows with famous people – just like it is in the movies and on TV.

Well, there is some of this, but the average reporter spends more time checking the city and phone directories on how to spell a name or to get an address than he does rescuing people from burning buildings, and covers more dull meetings than major murders.

This, however, is a fairly well known fact of life.

What is less well known is that the reporter will spend much of his life just waiting.

To be a good reporter, you must learn how to sit or stand indefinitely and patiently.

This is not fun, it is not exciting, it is not glamorous or romantic. It is hard on your feet or your bottom, depending on whether you are standing or sitting, and it is harder on your nerves.

I think one reason I am so prejudiced against waiting my turn for a doctor, barber, or dentist, or standing in line to see a movie or to get into a cafeteria is that I have had to wait so much as part of my job.

As a private citizen, you can get out of a line if it is too long and you can stalk out of an office if you must wait too long for an appointment.

But when you're a reporter and have to wait you can't do anything about it, except gripe.

I was thinking about this the other day when I spent about two hours sitting on a hard bench outside the state attorney's office, waiting for a press conference which turned out to be the announcement that three suspects in a terror bombing ring had been arrested.

It was a good story and worth waiting for, but while we reporters were waiting we didn't know that for sure. We only knew that it was our job to sit or stand – and wait.

A couple of weeks ago I was one of many reporters assigned to meet the plane bringing wounded and sick survivors of the Cuban invasion to Miami International Airport. The plane, of course, was late. Again it was the old, standard, routine of waiting – just waiting.

This is true so often that it truly is a routine part of a reporter's life. You sometimes feel as if you spend half your working hours sitting and standing, waiting for something to happen.

You may be waiting for the jury to return its verdict; you may be waiting for an important closed conference to adjourn so you can question the participants; you may be waiting for a person to come home or to his office; you may be waiting for a rocket or a man-carrying space ship to be launched; you may be waiting for the space ship to land.

But you wait. You have to wait. You think of the dinner you didn't have time to eat because you rushed to start waiting, or you think of how good a long, cool drink would taste, but you can't leave. You know that the second you leave that bench or that room or whatever place you are stationed that the story will break.

You may have sat there for 10 hours, but the second you go out for a cup of coffee or even visit the rest room, that's when things happen and you've wasted all your time and get bawled out by the city desk.

And that's why you develop patience. It may be an impatient type of patience, you may be restless and sore, both both in body and spirit, and you may say nasty things about your boss, your profession, and the people or elements that are holding up the story, but you sit or stand – and you wait.

It is the most trying and the most important part of a newspaperman's life, that waiting, and if journalism schools want to justify their existence, they ought to teach people how to do it.

Frankly, the only thing I can suggest for their course on this vital subject is to advise embryonic reporters to carry a pillow with them at all times. It makes sitting much easier.

A World at Breakfast

I have to have my newspaper every morning. I am lost if I can't have the news with my coffee.

Nothing can replace a newspaper. It doesn't have any serious competition, because it is unique.

I enjoy news broadcasts on TV or radio and I particularly enjoy pictures taken on the scene of important or exciting events.

But the next morning I still like the world to be delivered to my door, neatly wrapped up in a package, ready to be digested at my leisure.

Even if I am on the scene of a story, I still like to read the complete account of what I saw, as gathered from many witnesses.

Even if I see a play, hear a concert or watch a ball game, I like to read about it the next morning.

Newspapers can't match TV or radio in bringing bulletins into the home at the moment something happens, or shortly after.

But TV or radio or the movies or nothing else can do the many things that newspapers do.

You may miss a broadcast, but your paper is there on your doorstep, waiting for you to read it.

In the excitement of an event, a commentator may miss much of what was going on, or the event itself may be incomplete, or its consequences confusing or misunderstood. But the newspaper assembles the facts and presents them completely and in focus, to lay them before you for your careful perusal.

This is not to mean that a newspaper doesn't assemble and deliver the news rapidly. But the days when a split second was all-important to the newspaper is past.

In those days, the first word of the news was through the papers. Now it comes out of a box or from a screen.

The newspaper's primary function no longer is to be the fastest with the news. It's primary function now is to be complete, accurate, and fair.

As secondary functions, it is entertaining and educational. It does all of its jobs with amazing success.

You don't tune in on the middle of a newspaper. It's all there. And you don't misunderstand what it says. It's there in cold type.

The fact that it can't disown errors is one reason why the newspaper is so conscientious in trying to avoid them.

Some newspapers are better, bigger, more sparkling or more scholarly than others. But with very rare exceptions, newspapers are honest mirrors of the world, trustworthy reporters on life, dependable messengers on current events, and respected members of the reader's family.

A newspaper is a versatile character of many talents and many facets. One edition of a paper can be all things to all people.

The man of the house may curse the leading editorial, but his wife will prepare his dinner from recipes in the same issue.

Grandmother may scoff at the psychology clinic of Dr. Crane, but Grandfather will faithfully follow the health advice of Dr. Molner.

The teen-aged daughter may think her brother wastes time by reading the sports page, but she faithfully follows Ann Landers or the teen-age panels.

Maybe Junior can't share Sister's enthusiasm for Peanuts, but he loves Dennis the Menace.

Maybe the Man of the House blows a gasket at the frothy offerings of the society columnists, but he can't understand why The Woman of the House doesn't appreciate the business and real estate writers.

A visitor from the Big City may scoff at the local commentators, but he finds an old friend in Walter Winchell.

Maybe he can't stand Westbrook Pegler or Drew Pearson, but he may be a rabid fan of Jim Bishop or George Sokolsky.

"Newspapers print nothing but sex and crime!" the preacher may thunder, but he turns to the daily offering by Billy Graham, or to the listings of church services, or the comprehensive stories of religious activities, from synagogues to cathedrals.

"I like the paper because it gives complete reports on stocks and bonds," says one man. "I like it because it gives the daily racing form," says another. "I like it because of the horoscope." Or "I like it for the crossword puzzle." Or I like it because of the daily listings of the movies, TV and radio programs." Or "I love the Glamor Clinic."

There is a column on bridge playing, a column on boating. There is a fishing editor and a bowling editor, a garden editor and a church editor, a real estate editor and a business editor, book, music, art and school editors.

There is news from wire services and from staff writers – important news and humorous features.

There is something for everyone in the newspaper. There are ads and there are stories. There are pictures and there are puzzles. The newspaper is as versatile as mankind, as big as the world, as interesting as the people who make up the world.

20,000 Hearts Cry Welcome As 60 Casualties Return

Sixty scarred warriors of the Cuban invasion returned to Miami Saturday for reunion with their families before a sober, silent crowd of 20,000 refugee countrymen.

This was not a riotous victory celebration, or a joyful homecoming. It was the humble return of proud men being ransomed on a pledge to pay two and a half million dollars.

By their slings and their stumps, their crutches and canes, bandages on an eye or cast on a foot, and most of all by their faces, these men showed what three days of war and a year in prison can do.

Still in Cuba are 1,119 other captives of that April 1961 invasion. The Cuban government has agreed to free them, too, for an additional 59.5 million dollars in cash.

Cuban Families Committee spokesmen who accompanied the wounded and sick men from Havana Saturday expressed hope that the money can be raised and the other captives freed in a group.

The liberated survivors said that until their comrades are ransomed, they still consider themselves prisoners and will return to Cuba to serve their terms if the money isn't paid.

At the prison in Havana, there was a brief demonstration as some of the prisoners balked at leaving their comrades behind.

For this and other, unexplained reasons the plane was more than two hours late taking off. Instead of the scheduled 10 a.m. arrival, it landed at 12:13 p.m.

The huge crowd – largest ever gathered at Miami International Airport – was patient in waiting, even as a hot sun melted early morning clouds and beat down unmercifully.

They had started to arrive by 7 a.m.

At first only 1,000 were going to be admitted to the southern observation deck which overlooks the PAA unloading zone. When it became obvious that this would not accommodate the would-be spectators, another 1,000 were allowed on the northern observation deck.

But the people kept coming and by 10 a.m., at least 2,000 were packed on each deck. Speaking in Spanish, announcers repeatedly urged them not to lean or push against the guard fence for fear the continued pressure might cause it to collapse.

And still the people came. Unable to get on the observation decks, they stood in the huge lobby of the terminal. The parking lots outside were filled to overflowing.

The lots have a capacity of 5,000 cars, and every place was taken, as well as much of the space between parking rows.

It was, said Alan Stewart, director of the Dade County Port Authority, the biggest crowd at the terminal since it was finished in 1959. He made the 20,000 estimate.

As the crowd grew and the wait lengthened, Metro police worried whether their 50-man detail would be sufficient to keep control. Fire trucks were called and stationed near the landing strip, in case hoses were needed.

To prevent the thousands of people on the observation deck from suddenly pushing into the already-crowded lobby as the men disembarked, it was announced their cars would "parade" around the front of the terminal before heading for Mercy Hospital.

This had the desired effect of drawing thousands of people from the terminal to the outside, where they waited to see the men drive by.

At 11:10 a.m., it was announced that Pan American Clipper Flight 422 had left Havana. This was the plane.

At 12:08 p.m. it was announced that all the men had been able to board the plane under their own power and that no stretchers or wheelchairs would be needed.

Fourteen men in dark trousers, white shirts and dark ties then marched on the field. All were escapees of the Bay of Pigs invasion. Three of them had been wounded. They were the color guard, carrying the flags of the United States, Cuba and the yellow and blue banner of Brigade Asalto 2506, the assault brigade.

And then the plane came in. Handkerchiefs were waved so that the observation decks looked as if hordes of sea gulls were fluttering in front, but there was no sound from the people.

Immigration and customs officials boarded the plane. Through the loud speaker came the whistling movie theme song of British prisoners in "The Bridge of the River Kwai." This "victory in defeat" song was played over and over, the background music for the drama that followed.

After a brief interval, the first of the 60 stood framed in the plane's doorway.

He was Noleo Montero Diaz, commander of the invasion brigades 3rd battalion.

He stopped at stiff attention as he saw the Cuban flag displayed from the south observation deck, saluted slowly then came down the steps.

Mayor Robert King High stepped forward to extend greetings.

Then there was Mrs. Montero running to him, and his two children clutching at his legs, and he and his family marched slowly through the line formed by his former comrades with the color guard, past the relatives of others yet to come from the plane, to the line of parked limousines which were to carry them to the hospital for thorough examinations.

After that, they came in a steady march.

Each man, wearing faded yellow sports shirts, stopped to salute the flag of the homeland he had tried to free. Some had to salute with the left hand, for the right arm was gone. Some could not stand at attention, for they had only one leg; one had none.

There was Rene de Lamar Jr., wounded in an arm, his big family there to meet him, from his 73-year-old father in a wheelchair to Louise, his four-month-old niece.

Juan Figueres saw for the first time his infant daughter, but he could not carry her, for he had only one leg to walk on.

And Lorenzo Porto Valdes, shot through the hip, was using a crutch.

Carlos Dosal embraced his wife and patted the heads of their three children with only one arm, and Adelberto Sanchez walked straight though he still bore a bullet in his chest.

The 75-yard walk to the limousines was long for some, particularly the young man who came home with two artificial legs, and for a few older men who walked slowly and with blank faces.

Almost all were met by some loved ones. The women held up the babies and the smaller children, and kissed their men, and almost all of them, burying their face against a shoulder, and sobbing as they walked to the limousines.

Almost without exception, the men received these greetings with slight show of emotion, although they put a reassuring arm about their wives. It was almost as if they were spectators, rather than participants.

Later, during the 15 minute wait before the men were put in the limousines, they began to relax. Finally, they smiled more freely and began to join in the babble of Spanish flowing from their excited relatives.

And through it all, the crowd watched in such a subdued manner it was almost a reverent silence.

Outdoors, in front of the terminal, as the limousines carried the men past the terminal and through the parking lots, it was different.

Then, the Cubans cheered and many of the men waved as they slowly went by.

And the cars honked and people cheered as they were driven along LeJeune Rd. behind police escorts.

There were six ambulances at the field, but only one was used. It took a wounded man, suffering from shock, to Jackson Memorial Hospital. Others were taken to Mercy, where an entire wing of the hospital was turned over to them.

Thirty-five of them were released, although some must return next week for more checkups or treatment. The others were admitted to the hospital.

Most of the sixty men who came back from Cuba spent Saturday night with their families.

But they still were not home.

'Twas Truly Dramatic

The reporters and photographers, the TV and radio men who covered the arrival of 60 wounded and sick survivors of the 1961 Cuban invasion had expected something much different.

We expected a frenzied, wild reception, an exuberant, hysterical demonstration. Knowing how excited a Cuban gets in ordinary conversation, we figured that when thousands of them packed together in a monster mass let loose of their emotions simultaneously that pandemonium nearing riot would result.

Officials shared that view. And as the crowd grew and grew, far beyond the wildest expectations, police and port authority officials began to worry about whether they could control this enormous, emotional crowd when the plane arrived.

There was no need to worry.

Never was there a more orderly crowd. There was no mass hysteria, no mass effort to break through police lines, no demonstration which could have become uncontrollable.

It was a dramatic, touching, poignant scene as thee wounded and the sick men came from the plane. But there was no violent outburst by the people. Instead they were quiet, acting not as a delirious mob but as solemn spectators of a moving drama.

There were many reasons for the good conduct of the huge crowd.

The Metro police, port authority and the Cuban committee which obtained release of the prisoners from Castro had planned well.

Every precaution to keep the crowd from getting out of control had been taken. The years of experience of county police in handling airport crowds was shown by the calm and efficient manner they handled this situation.

But the main reason for the restrained behavior of the people was that there was no reason for any other behavior.

We gentlemen of the press and the officers were thinking of these freed prisoners in terms of American soldiers coming home from a war and imprisonment and we were thinking of the crowd in terms of an American crowd welcoming their soldier heroes home.

But these men were not coming home. Cuba is their home. They had been here as refugees and they had gone home in an invasion against a dictator they despised. And they had lost. It had not been their fault, but they knew that their invasion had been a tragic farce almost without precedent.

They knew, too, that they had been bought out of prison, that the dictator they tried to depose was strengthening his hold on their homeland by the tribute paid for their release, so that by their return they actually were helping support their enemy.

Even more, they knew that they were only a portion of the prisoners of that ill-fated invasion, and they had a sense of guilt that they were free while their comrades remained in prison, to stay there until they, too, may be ransomed.

They were proud patriots still, these men, but certainly they did not feel like heroes returning triumphantly home.

The relatives waiting for them were in no festive mood, either. True, they were overjoyed to be reunited with their men. But they were not greeting the same men they had last seen more than a year ago.

Those were confident, carefree, laughing young men, filled with the spirit of adventure and the courage of a crusade.

These men who came back were the beaten, the battered, the sick, the lame, the blind, the armless, the living casualties of a lost battle.

And the other thousands of Cubans three to greet them – they were solemn because they were witnesses to all this, and because they were thinking of their own sons, husbands, and brothers, thinking as these men who came back were thinking, of the 1,119 prisoners still in Principe Prison.

That was the main reason for the quiet conduct of the crowd at the airport.

We Americans had misunderstood when we thought of the occasion as a celebration or a homecoming.

It was not a celebration. It was a chapter in tragedy.

Incidentally, never have I seen better cooperation between police and press under more difficult circumstances than at the airport that day.

As far as I know, there was not one complaint by police about the press or one complaint by the press about the police, and none against either by the people.

Larry's Drive for Trees

Finest Season's Now

Florida has good seasons (comparatively anyway) around the calendar, but if I had to pick a favorite, I would choose right now. I am speaking of that period in spring in late April and early May when the days are clear, the clouds white, the sun warm, and there's enough of a breeze to fill a sail or air condition the nights. And it is the season when the flowering trees are at their best.

A drive around Greater Miami – and I am sure this must be true in most areas – will show you trees waving pink, purple, blue, white, and all shades of red and yellow blossoms in the air – brilliant additions to the scenery.

Even the royal poincianas are beginning to raise their orange-red umbrellas, although the eccentric tree at our house is a hold out, demanding more fertilizer and shorter blooming seasons. I am confident that it will eventually join the others.

Right now is when the Sunshine State Parkway would be the talk of the tourists and the most attractive highway in the nation if it were lined with flowering trees.

Some day, if we live long enough, we may see that Boulevard of Blooms become a reality.

Big Trees Require Time

If orchid trees were ordered a year ago, it seems strange that they just now are being put out along the Florida turnpike, but there seems to be a valid explanation.

Mighty oaks from little acorns grow, but it isn't an overnight process. Like Rome, flowering trees don't happen in a day.

The way it was explained by B. M. Brown, landscape superintendent for the turnpike, I think I understand how the tree-planting program works.

The turnpike authority operates nurseries at Fort Pierce and Kissimmee. Nearly everything is grown from seeds or cuttings. They are started in "mist beds," where they are sprayed with a fine mist for several seconds out of every minute.

When the roots are strong enough, they are moved into gallon cans. They are watered by an automatic sprinkler system. In the case of trees, they later are moved into larger containers and when they are big enough are transplanted to permanent sites.

Even then, the tree won't look like much when it is put out, and those trees already planted along several plazas and interchanges probably look more like Twiggy than a be-jeweled Mae West. But within a year most of them will bloom, although some take longer.

However, Mr. Brown promises that by next spring a driver along the Sunshine Parkway can tell that roadside trees are waving flowers at him. That's what I'm anxious to see.

If the program is carried on the way it should be, the Sunshine Parkway flowers will be as well known as the cherry blossoms in Washington, the roses in Portland or the azaleas in Mobile. Maybe people will notice them almost as much as they do our pretty girls, although that is expecting a lot.

Nature's a Tough Gal

Written on a dreary afternoon in Miami last week, just before a storm that didn't turn out so bad after all.

The sky is like lead, dull and heavy, and it presses down against the earth, squeezing the life out of the air. It is quiet, ominously still, and nothing moves in the motionless air.

There is no sun, only the leaden, oppressive sky, and if there are birds about they are not singing, and if there are people about, they are scurrying furtively, and they glance apprehensively at the brooding sky, as if hoping they can reach protection before it spills its wrath upon them.

This is Nature when it has declared war upon man, and the elements are holding their breath, waiting the moment when they will unleash their vengeance on the people.

And puny man realizes his insignificance as he waits to see what Nature will do to him.

We have tampered with Nature, we have at times harnessed and sometimes controlled Nature, but we have never conquered or enslaved Nature, and we well know that when she becomes mad that we are helpless before her.

We have diverted her streams and dammed her rivers, we have created energy from her waterfalls and taken energy from her sun; we have split atoms and we have bred germs to fight other germs.

We have sailed under her seas and soared above her skies, we have broken her rocks and drained her veins, all for our own uses. We have made machines that fly higher and faster than birds, machines that follow tracks through mountains at a speed no animal could equal, we have built machines so huge that we load them with hundreds of our humanity and so efficient we send them faster than the speed of sound far above the atmosphere.

But we know that if Nature becomes angry, that our strongest moving machines will be tossed about by the elements like match sticks and fall apart like a child's toy in the grip of a giant.

We have cut down trees and drained swamps and we have poured ribbons of concrete across our lands and we have machines that carry the power of hundreds of horses under their hoods as they take our humanity over torrents and through granite peaks.

But we know that Nature can flood our highways and render them useless, that floods can tear away our bridges and rock slides can close our tunnels.

We know that the elements can make our machines useless if the elements become furious.

By earthquake or flood, hurricane or tornado, drought or tidal wave, glaciers, dust storms or blizzards, lava pouring from the earth or water from the heavens, by any of the many means at its command. Nature can halt our transportation, disrupt our communcationsi and make of our proud sky-scrapers and split-level residences mere caves in which we cower and hope for deliverance.

If the sun comes too close, it will destroy us, and if the glaciers invade our sunny lands they can destroy us, or if the oceans spread over our continents they will destroy us, or if it rains without end or if it does not rain without end, we can be destroyed.

That is the message of the leaden sky and the deathly silence before the storm. That is the message to the man who has challenged Nature and oft times believes he has mastered it, and that is the message which, not quite comprehending, puny man receives as he gazes apprehensively at the heavy sky which is closing in on him.

The storms pass away, though they need not always, and puny man, who can be majestic in times of travail, usually builds back that which Nature has destroyed. And much of his majesty in such times of rebuiding comes from the humbleness which he has learned, but which he so easily forgets.

But in the quiet before the storm, mankind receives the warning, the reminder, which it needs on occasion lest it be carried away by its arrogance.

The Day Must End

Somebody must write the obituaries, but it is not a pleasant task . . . especially if you are writing about somebody who was a friend . . .

There were many times, when I was young, when I would look about me and see the many people . . . Always there were so many people . . . Vital, virile, exuberant, healthy people . . . And it was difficult for me to believe in Death . . . It was difficult for me to believe that all of these so-much-alive people some day would be gone . . . I knew that people of past generations had died . . . but it seemed impossible that all of those who now peopled the earth would one day be gone ...

In any place, among any people, the hand of Death will be felt . . . Although in your younger years, those who are taken may be so much older, so little involved in your intimate life, that you still can't understand the inexorable strength of its touch . . .

When I came to The Herald and made friends among my fellow workers, I little thought, as I worked with them and laughed with them, that many would one day be a subject for an obituary in my typewriter . . .

The first was Henning Heldt, a good friend and splendid reporter . . . and in the following years there have been stories on the lives and deaths of such Herald friends as Al Hallman, Abe Hurwitz, Mary Schuck Smith, Ernie Hill, Lee Gash, Ferman Wilson, Ed Pfister, John Kilgore, Jesse Irwin, Bob Hanson, Merlin Test . . . and now, Roy Kelly . . . And most of these obituaries I have written . . .

And somewhere in The Herald City room now there is a fellow reporter who some day will write about the life and death of Larry Thompson . . . If Charlie Ward thinks there's enough space in the paper for an obit that day . . . And, in time, all of the people here will end up in that final column . . .

But the typewriters will continue to clatter and the presses will continue to roll . . . And Henning Heldt's daughters are grown . . . and so is Ernie Hill's son . . . My oldest daughter's daughter soon will be two years old, and my youngest daughter starts school next year . . .

And the people who fill the earth will bear our names . . . And will live their lives from what they learned by the way we lived our lives . . . and the young of a future generation may look about them and see the virile, the vigorous, the vital people about them and doubt that there really is a Death that can remove them all from the earth . . . and they will be right.